DATE DUE

DEMCO 38-296

Meaning in Technology

Contents

Meaning in Technology

Arnold Pacey

The MIT Press
Cambridge, Massachusetts
London, England

©1999 Arnold Pacey

This book was set in Sabon by Wellington Graphics.

Printed and bound in the United States of America.

Library of Congress Cataloging-in-Publication Data

Pacey, Arnold.
  Meaning in technology / Arnold Pacey.
    p. cm.
  Includes biliographical references.
  ISBN 0-262-16182-6 (hc : alk. paper)
  1. Technology—Philosophy. 2. Technology—Social aspects.
  I. Title.
  T14.P28 1999
  601—dc21   98-49287
                                                        CIP

Acknowledgments

A critical problem admitted by Frederick Turner, in his splendidly stimulating book, *Beyond Geography* (1980; new edition 1992), was "how to write on a subject whose scope was far beyond my competence." So it has been with me, and hence the sincerity of my thanks to a wide variety of friends and colleagues—especially those who have given me the confidence to carry on, through many rewritings. It is also the case that my reading does not usually range as widely as appears here, and I am especially indebted to five people whose gifts of books and supplies of citations, offprints, and photocopies have widened and deepened my thinking, namely Penny Williams of Toronto, Carole Brooke of Lincoln, Gregory McIsaac of the University of Illinois, my brother Philip Pacey, and Jerry Ravetz. I also owe a special debt to Maggie Horsman, who runs the tiny village library in Addingham, for obtaining many of the more obscure books cited. In addition, Phil Gates provided a photocopy needed for a last-minute addition, and the BBC supplied the transcript of a radio program quoted in Chapter 7.

Several of the above have also read draft chapters, as have Michael Bartholomew and Jo Peach, both of the Open University, and also Howard Rosenbrock, Averil Stedeford, and Sara Wermiel. Two who have read and commented on the book as a whole have been especially helpful, namely my valued colleague, Denise Crook (also of the Open University), and again Philip Pacey.

The book has been about nine years in the writing, and ideas for it have been in my mind for much longer. Over that time, many more people have helped me or provided stimulus than I can now recall. But I do remember Penny and Peter Jones for the gardens of Chapter 6, and Alison

Armstrong for some visual thinking in Chapter 2. Chapter 7 owes much to my late mother's experience of midwifery, but I also explored some topics in this chapter and Chapter 9 in three lectures given at the University of Virginia in October 1991. For that opportunity, and for the chance to enlarge my understanding of Thomas Jefferson (see Chapter 6), I am indebted to Kathryn Neeley and Alan Gianniny.

It may seem strange that a book stressing visual thinking is not profusely illustrated. There were no funds to do this, but on a deeper level, the struggle to get ideas down in words seemed to inhibit visual imagination, thereby confirming the view expressed in Chapter 2 about the differences between verbal and visual modes of thought. However, I have to thank artist John Nellist for sharing ideas on the subject, and it is likely that any visual content in the book will owe something to him as well as to designers at MIT Press. For other help received in the final stages of the work, I am immensely grateful once again for Dinie Blake's invaluable work as a proofreader.

Last, but most important, thanks to my agent and publisher are fundamental and heartfelt. A dialogue extending over more than two decades with people associated with MIT Press, especially Larry Cohen, has made it possible to build from my first book in ways that I could not have anticipated. Two books were written as a result of prompting from that quarter, and this one has had much encouragement. So what might have been a trilogy now contains four volumes. Similarly, the practical support that has always been available from Lisa Eveleigh, my agent, has done much to prevent discouragement during the very long haul involved in completing this work.

Introduction: Dimensions of Experience

The Architecture of Progress

A formative moment during the nineteenth century was the Great Exhibition of All Nations held in London in 1851. It was a moment for reflection about the different ways competing nations were using science and invention. It was also the first in a series of "world's fairs" and trade fairs in which one city after another celebrated the meaning of technology in its most public sense. Especially notable were exhibitions in Paris (1867 and 1889), Philadelphia (1876) and Chicago (1894). In the twentieth century, Chicago and New York successively and competitively held world's fairs during the 1930s.

Much of what went on at these events could be regarded as the promotion of an ideology of progress and the advertising of national and corporate achievement. Beginning with the "Crystal Palace" that housed the 1851 exhibition, the best buildings constructed for world's fairs combined with the most striking exhibits to demonstrate how engineering and other technologies could be means of expressing human aspiration for the future, as well as celebrating the progress already achieved.

Achievement and aspiration, progress and purpose were all part of the public meaning of technology. Furthermore, the works of engineering displayed were increasingly seen as "sublime," and endowed with meanings once found only in the wonders of nature.[1] However, the public meanings of technology are not the only ones that matter. When the centenary of the first Great Exhibition was marked by another celebration of technology in London during 1951, I was of an age to respond with an enthusiasm that extended even to the science being taught at school. The word "technology" was hardly used at that time, or where it was used, it referred to an activity *led* by science. The geology exhibit I saw showed how "science is revealing the age and structure of the earth," while "technology develops . . . (the) underground resources . . ." thereby revealed. As I went around the exhibition, a young boy with his father, the word "science" seemed the central idea, linked directly with wonderful improvements in health and daily living.

At the time, Britain was still scarred by war damage. On the way to the exhibition we could see ruins in central London where buildings had been destroyed by German bombing. They were temporarily a source of wonder for the way nature was taking them over, and we saw them resplendent with the pinkish-red flowers of rosebay willow-herb. The relics of war lent the exhibition a more serious mood than some of its predecessors. There was a determination to use science to create a better world as the ruins were rebuilt. Moreover, the architecture of the exhibition, with its Dome of Discovery, its "skylon" spire, and the noble Festival Hall, seemed an exciting foretaste of the architectural forms rebuilding might use.

In retrospect, it seems that my strongest response was to the architecture, but this had a close relationship with what I felt about science, with all its potential to assist with rebuilding and renewal. Included in that was a degree of enthusiasm for nuclear physics. At the exhibition we were told about important medical applications. They were illustrated by reference to radioactive tracers, in a generally triumphant explanation of how "the horizons for human endeavour will continue to expand."[2]

Many others must have reacted in a similar way to the 1951 Festival of Britain (South Bank) Exhibition, whose architecture and design set a fashion, and whose message about science was so hopeful. This was all part of the public meaning of technology, but I tell it as a personal story in response to comments about "the meaning of technology for the way we live" put forward by Langdon Winner, one of the most perceptive writers on the subject. He notes, in particular, that there are certain questions which academic commentators on technology (and other subjects) do not often ask. A fear of subjectivity "leads many people to write with as little personal character or self-reference as possible." Personal experience, even when it is relevant to understanding public meanings, is "scrupulously avoided."[3]

One reason for this book is a belief that to understand technology, we occasionally need to acknowledge and be aware of personal experience. And although my own experience of technology in 1951 was not exceptional, there was certainly something individual in the way I responded to the architecture of the exhibition, as if that itself was an aspect of technology. This is still a bias in my approach, as may be evident in Chapter 2. For me, the lovely forms of an airliner or suspension bridge, the elegant conceptual structures of mathematics, and even talk of the "architecture" of computer systems may evoke positive visual responses. I have always wanted to be working in architecture while feeling that I *ought* to be doing technology or science, or writing books like this.

Twenty years after the 1951 exhibition, my first attempt at a serious exploration of technology, mainly through its history, took architecture as a model from the start. The book that resulted tried to define the goals of innovators, engineers and other technologists in terms of conflicts among different forms of *idealism*. Especially important, because it reflected conflicts in my own life, was the aesthetic idealism of architectural and engineering form, as contrasted with the social idealism expressed in technology applied directly to promote human well-being.[4] In a later study of the "culture" of technology, where architectural themes had a lesser role, I attempted to discuss some of the same conflicts on a more political level, though with reference to individual motivations as related to personal *values*.[5]

"Values" and "ideals" were almost interchangeable concepts for the early books, in discussion of links between technology practice and individual purpose and experience. However, in this third and final exploration, I prefer to use words such as "aspiration" and "meaning," believing that a person's ideals and values in relation to technology are an outcome of her or his sense of the purpose and meaning of life.

Personal Responses to Technology

Many commentators on technology regard discussion of ideals, values, or meaning in technology as futile because they are inherently subjective. Such people might acknowledge that ideals and imagination are part of an individual's experience and may affect that person's work in technology, but they say that ideals cannot be objectively observed. Imagination cannot be measured (except perhaps in psychological tests of dubious value). Instead, such commentators often find it more fruitful to seek to understand the wider role of technology in society by discussing the "political economy" of its development and use. This approach was especially illuminating in the 1970s and 1980s, for exploring contrasts between high technology and "alternative" or "appropriate" technologies,[6] or in bringing out the implications of the introduction of new technologies.[7] Today, though, another school of thought focuses on different technologies and brings out other implications. Case studies include the invention of the bicycle in the nineteenth century, the evolution of Bakelite, and a more recent electric car project. These innovations are seen as "socially constructed," in the sense that there is no one individual imagination behind their development, but instead a variety of "actors" responding to a complex of social pressures. Invention is thus seen as a process involving many people.[8]

This viewpoint contains some important insights, and I have learned a good deal by reading the work of its advocates, just as I have learned from those who discuss the political economy of technology. On some topics, notably related to the energy industries, it is also fruitful to notice conflicts between social constructivist and political interpretations.[9] Yet few authors of either school get close to what seems to me the most important aspect of the practice of technology, which must be related to how human minds work, and how individuals act. That may include reference to how people respond to social and political circumstances, of course, but it is also important to ask how human imagination deals with practical experience of the material world.

So although I am very ready to acknowledge that studies of social construction and the political economy of technology have led to important insights, I am concerned that these approaches are quite often linked to a refusal to acknowledge personal or imaginative responses to technology. The result is that experience of innovation, responses of consumers, and the education of students in technology are all too narrowly interpreted.

By contrast, one of the best short books to touch on the themes I tackle here is the work of an educationist, John Head. He has taught science in both the United States and Britain and has studied the theory of education. On the basis of that experience, he defines a central issue as "the personal response to science" by individual students. He notes that although learning science is intellectually challenging and therefore makes heavy cognitive demands, "the possibility that there might be . . . affective demands may be less obvious."[10] Yet there are such demands, which relate to the individual's motivations in studying science, and which interact with what qualities that person most values in nature and in other people (as we shall see in the second part of this book).

John Head identifies some personal responses to science, and some perceptions of technology, in which male and female students tend to differ markedly. Among other groups, not just students, gender appears to be a very significant factor in attitudes toward technology (and to related ethical problems) as we will find in the second part of the book. However, differences in experience between consumers and engineers, designers and craftworkers are also significant, as are other contrasts in personal background and temperament. For example, we might compare reactions to various applications of computing and discover that people "respond quite individually to technologies and feel comfortable with them in different ways and for different purposes." Somebody who thinks visually (in a sense explained in Chapter 2) might be excited by manipulating a drawn image on a computer screen while feeling uncomfortable using a computer as a word processor.[11]

One way of investigating differences of this kind between individuals is through psychology, which is how John Head tackles his discussion of the personal response to science. In the present book, also, the work of psychologists is quoted quite often, and on one level, it is reasonable to think of some chapters as being concerned with the psychology of technology.

However, psychology tends to explain human behavior and experience in terms of mental processes, and what causes them to operate in one way rather than another. By contrast, I am inclined to affirm the value of experience in its own right, just as experience, rather than regarding it as needing to be reduced to some basic explanatory scheme. That is to say, my standpoint is *affirmative* rather than *reductionist*. I wish to ask what it feels like to practice engineering, or to use a machine. Therefore, in the first half of this book, I seek to describe the visual, tactile, and indeed, musical experience that accompanies various kinds of technological and mathematical work, even while taking note of what psychologists say to explain it.

Against Reductionism

If psychology is of only limited value for discussing subjective experience, another approach is to ask whether this kind of experience might be seen as another *level* of knowledge. Some commentators suggest that beneath the public knowledge of technology and science that is set out in textbooks and professional discourse, another kind of knowledge is operating in a less explicit way. John Head, for example, says that we need "a clear distinction between the public knowledge of science and the individual's personal understanding." Others talk about the unspoken "tacit knowledge" involved in practical work, and Michael Polanyi includes this in his discussion of "personal knowledge."[12]

With a different emphasis, Gerald Holton makes an instructive distinction between the private knowledge of leading scientists and the formal statements in their published work. Private thinking can follow surprising directions in producing new ideas. Yet results are described and justified according to the norms of scientific discourse in its public form.[13]

This leads me to think of different interpretations of technology as belonging in a hierarchy of levels, with the politics of technology concerned with the most public and general level. On another, still public level are the published writings of engineers and applied scientists, and the efforts of histories and social constructivist commentators to interpret them. Beyond that, it is possible to focus down to the level of the individual and ask about his or her private experience of technology. This might lead to an analytical study like Holton's, or a discussion of psychology in biographies of technologists, or an account of the "pleasures of engineering" such as Samuel Florman provides.[14] Social scientists sometimes try to investigate experience at this level through questionnaires and interviews, and there is classic work of that sort by Anne Roe (discussed at length in Chapter 2), and by Mitroff and colleagues.[15]

The latter authors are especially interested in the supportive role a spouse may have in relation to his or her partner's scientific work. Their study is of considerable interest with regard to gender relations, but seems only to touch the fringes of personal experience because its social science methodology operates in terms of what can be objectively investigated. As philosopher Mary Midgley says: "The academic's dream of pure sanitized objectivity only leads us to conceal essential material." My aim here is to seek understanding on a more personal, more inward level that takes account of feeling and imagination. That seems an effort worth making because, as Midgley argues, unless we appreciate the significance of people's inner lives, concepts such as *creativity, will, purpose,* and *ethical responsibility* become difficult to handle.[16]

However, Midgley dislikes the tendency to talk about these matters as belonging to a different level of experience or meaning. Her objection is that once one uses this kind of language (which I do only tentatively, and for convenience), there may be an implication that some levels are more fundamental than others.

Another way of thinking about this may be to regard all activities involved in practicing or using technology as having several dimensions. In a previous book, I used a triangular diagram to suggest how "technology practice" not only involves hardware, practical skills and technical knowledge, but also involves an organizational, political dimension, and a "cultural" aspect relating to values and beliefs (figure 1).[17]

That diagram is not adequate for this book, as the view I am suggesting here implies that general talk about "cultural values" is not enough. Rather, we need to distinguish personal values and individual experience of technology from shared, social meanings. This calls for a perspective drawing to indicate a three-dimensional model (as shown on page 8). Here, the personal dimension is shown below the others merely to indicate that it is often hidden. Thus, although different levels of meaning are occasionally spoken of in later chapters, this diagram may serve as a reminder that relationships between the various meanings in technology can be represented in a variety of different ways.

Regarding the somewhat different topics on which Mary Midgley writes, her concern in stressing that no one level is more basic or fundamental than others is directed especially to scientists and philosophers who hold reductionist beliefs, and who assume that there is some basic level of knowledge that provides ultimate explanations. For biologists,

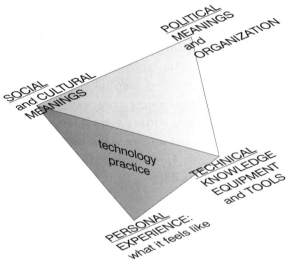

Figure 1
Dimensions of technology practice and experience

genetics is an area of fundamental knowledge that, they sometimes imply, can explain observations made at other, less fundamental levels, such as during studies of animal behaviour. Even the behaviour of people is claimed to be "the circuitous technique by which human genetic material has been and will be kept intact."[18] Similarly, it is often said that physics is a fundamental discipline that can provide basic explanations to which complex phenomena studied by other scientists can be "reduced."

Midgley argues, though, that it is better to regard different kinds of knowledge as coming from different viewpoints, and these ought *not* to to be arranged in a hierarchy. Rather, one should use whatever kind of knowledge is most applicable and relevant. She points out that not only are "fundamental" sciences sometimes given "despotic" status, but so are ideological systems and religions. In exploring the various meanings to be found in technology, I have encountered a writer who claims that *only* by adopting a psychoanalytic view will we discover what technology is really about.[19] That is to treat psychoanalysis as if it had despotic status. Similarly, there are writers on political or social meanings in technology who see their analyses as unique sources of insight, and say that *only* by understanding power relations (or perhaps social constructivism) will we learn.

But there is no "only" about it. We need to use political economy to understand some problems, and psychoanalytical insights to understand others. There should be no reductionist scheme for dealing with everything in terms of a single mode of explanation. The dynamic of technological change seems to reflect a synergy among psychological, institutional, and larger socioeconomic movements. It is a mistake to assume that any one part of this complex interaction is the key to all of it.

Human Purposes

It is not my aim, therefore, to construct arguments critical of views that differ from mine (except for reductionist views), nor against views that seem incomplete. The purpose of this book, rather, is to explore my own vision of the meaning of technology and some related sciences, knowing that this vision is partial and needs to be complemented by others.

One illustration of my view is provided by the earlier point about "tacit knowledge." Such knowledge includes unstated assumptions, skills that can be applied without thinking, and inarticulate visual awareness. In the prescientific past, craft workers often developed sophisticated skills based almost wholly on knowledge of this sort, and today, engineers and scientists may do things on the basis of hunches they cannot easily explain. Recent work on robots, expert systems, and artificial intelligence has attempted to tease out the detail of a wide range of manual and intellectual skills, including ones apparently based on tacit knowledge. It has been claimed that, no matter how deeply embedded within unarticulated experience a skill may be, the knowledge it represents can be made explicit and objective by being built into machines. But despite some remarkable successes, initial optimism about expert systems (for example) has been considerably blunted, and they are now seen as relevant mainly where strict rules and logical procedures apply and where context-based insight is unimportant. They are best at the chess-playing style of problem, for example. It is also worth pointing out that tacit knowledge in a human mind not only is operational knowledge that can easily be built into a machine, but also includes a sense of what the knowledge means and how it is related to human purposes. That kind of "context-based insight" is partly what this book attempts to discuss.

A second illustration arises from John Head's comment that it was once a common assumption in psychology that human minds would be "inert and passive" in the absence of external stimulus. They were only "provoked into action by some internal or external force," and psychologists of different schools differed as to what were the most powerful forces motivating inert minds: "thirst, hunger, sexual attraction or maternal care."[20]

This is the cause-and-effect model of human behavior. Hunger or the sex drive is supposed to "cause" activity such as hunting or the search for a mate. But as Head notes, more recent work in psychology has suggested that this may be partly misleading, or even incorrect. Observation of animals shows that they are often highly active without any obvious stimulus, and this activity has to be "unsatisfactorily labelled 'play' or attributed to 'curiosity'." As an educationist, Head goes on to note that children "do not need to be motivated to learn." Until they are sent to school and asked to learn things that do not interest them, learning for children "seems to be a spontaneously occurring activity."

Words like "spontaneous activity" and "play" suggest random and undirected activity. But that may not be the right way of looking at it. At issue is that in contrast to the old psychologists' model of stimuli causing activity, there is now recognition that some actions come from *within* a child (or older person). Often, that means actions directed by a purpose rather than a cause. There is a metaphysical minefield surrounding ideas about purpose, but our experience as we grow from childhood is that we gradually become more aware of our own purposiveness. Then we endow work and sport, hobbies and adult play, with deliberately formulated purposes. There may be important differences between play in children and adult thinking of this sort, but there seems to be continuity between the two as a person grows to maturity.

These thoughts may shed light on the points made earlier about tacit knowledge. Where technical skills are at issue, tacit knowledge is not just knowledge of materials and how they can be worked, all of which might be captured by a computer system and robot. It may also include experience of affective responses to technology, concerned with purpose, aspiration, and relationships. All these are topics of importance for understanding meaning in technology.

Because my approach to this subject is influenced by knowledge of the history of technology, it is also worth saying that my interest in history centers on how human purposes, aspirations, and relationships work themselves out in technological contexts over time. The contrast is with historians more interested in "causal models" who then engage in (to my mind) fruitless debate about "the causes" of industrial revolution—or they ask what "determines" social change and whether there is a process of "technological determinism" at work.[21] But in history, as I see it, technology is primarily an expression of varied, often confused human purposes, and these should be the center of interest.

Aims of the Book

In discussing experience of technology as it is encountered by individuals—engineers and mathematicians, craft workers and consumers, women and men—my aim in this book is first the simple one of affirming that personal experience is significant, and something we ought to acknowledge.

My second aim, though, is to consider ways of discussing individual experience that avoid devaluing it with comments about the "merely subjective." In seeking a framework suitable for this purpose, I reluctantly borrow a simplified form of some psychologists' vocabulary in places, and I also talk about levels of experience or dimensions of practice, as explained above. However, there is still a question of whether it might be better to talk about ideals or values (as I have done previously), or about personal responses (as in John Head's educational work).

The organization of these ideas is represented partly by dividing the book into two parts, one dealing with direct experience of technology, and the other with contexts in which technology is used, relating to nature and society. Some of Head's language about affective responses seems useful in discussing personal relationships and gender issues as they interact with technology, but the title is *Meaning in Technology* because I believe that experience of the scientific and technical interacts strongly with our sense that life itself has meaning, and that there is purpose in living, even if we cannot define it.

Third, in dealing with these issues, I am inevitably probing the wider philosophy of technology and the assumptions about the world and about nature on which it is based. That leads to questions about the basic models or paradigms we use when discussing what technology is about. I do not pretend to say anything profound on this more demanding topic, but I ask three kinds of questions about paradigms.

To begin with, can paradigms for technology reflect human experience of purpose as well as scientific experience of causality (Chapters 1 and 2)? Next, how do paradigms for technology reflect the relationship of humans to nature (Chapters 3, 5 and 6)? To amplify this second question, are humans seen as detached from nature and as having dominion over it as managers? Or are humans understood as part of nature—as participants—and hence as needing to tailor their ambitions to what nature can accommodate? Finally, how do the paradigms we use portray the roles of humans relative to their technology (Chapters 7–9)? Are humans seen as detached, and hence outside their technological systems? If so, does it follow that the ideal technological system is one that is so perfectly automatic that it can operate without any people around at all, as a little "world without people"? Or are humans seen as participants in technology, even of the most advanced kind, and would it make sense to aim for the creation of a people-centered or "human-centered"[22] technology?

An important distinction in all these comparisons seems to be whether a *detached* way of thinking is possible in all kinds of work in technology and science, or whether the reality often is that we adopt a *participatory* approach, feeling ourselves to be involved in the system on which we are working. In stressing limitations and dangers associated with detached ways of thinking, and emphasizing insights arising from intuitive, participatory approaches, I will undoubtedly be accused of sentimentality, or of elevating feeling above reason, or of wishing to be "emotionally correct." But I will be dismayed if this book is regarded as contributing to the current fashion for abandoning serious thought and debate in favor only of having the right feelings about events. My argument is that we should acknowledge the existence of feelings, not that we should indulge them.

An important lesson to learn from creative work in science and engineering is that although ideas may arise in all sorts of ways that may be described as intuitive or participatory, there is always an obligation to translate them into more rigorous, often mathematical formulations, so that others may understand and check them, and explore their precise implications. It is a mistake to ignore intuition and personal feeling in a pretense of being detached and objective. But it is dangerous and contemptible to elevate the importance of feeling while ignoring the obligation to be as rigorous, logical, and rational as is humanly possible in checking out the direction in which feelings may be pointing.

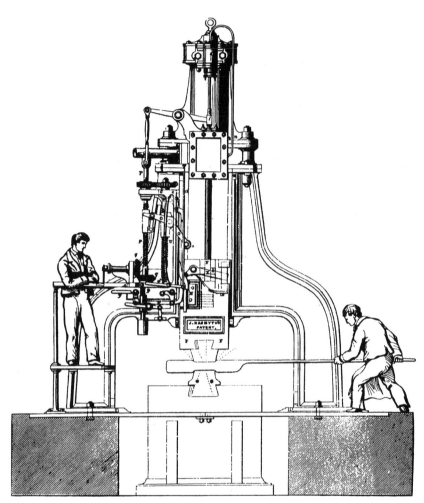

Figure 2
James Nasmyth's steam hammer, invented with the aid of "visual thinking,"
1839 (from W. Fairbairn, *Iron: Its History, Properties, and Processes of Manufacture,* Adam and Charles Black, 1861, p. 120)

I

The Practitioner's Experience: Visual and Musical Fundamentals

1

Music, Source of Technology?

Singing the World

The aboriginal people of Australia believed that the world began only when their Ancestor sang it into existence.[1] Many others can feel the sense in this, because singing (and other music making) is so potent a way of finding meaning in life. Rebecca West, in a novel, has somebody asserting that one need not be defeated by an incurable cancer when "there's all this music in the world," because music is "about life . . . and specially about the parts of life we do not understand."[2] And at the end of the twentieth century, music brings a spiritual dimension to the lives of the many people who no longer practice any religion. Thus although there may have been a world of sorts before humans sang about it, we may well feel that it did not mean so much. Although there was life, it could not be confident of meaning in the face of death.

Singing also implies dancing and poetry. And we make music not only with voices but also with instruments and machines. That is where the theme of this book begins. My view is that we sometimes—perhaps increasingly—use machines and other technology in the same way as we use music and musical instruments, to interpret the world and give it meaning. Often, this involves conceptual or visual aspects of technology, but at times, it is the sounds made by machines that convey meaning. Nor is that just a matter of musical instruments that happen to be quite complex mechanisms, such as pianos must have seemed when first introduced, or the drum machines and synthesizers of today. It has more to do with the way some machines intended to serve more mundane functions take on a musical role. When my bicycle is going well and I have an empty road, I find myself singing to the quiet rhythms of its motion. The clickety-clack of power looms in cotton mills made Lancashire clog dancers want to tap-dance the pattern of their rhythms. Church organists are often strongly drawn to steam engines, which they seem to recognize as a related form of instrument. The motion of ships has contributed to the lilt of sea shanties, and of trains to boogie-woogie (and more formal music too, notably by Honegger and Prokofiev). Murray Schafer has commented that the internal combustion engine gave music a new note, and this is appreciated, at least, by young men with motorcycles who adjust their machines and ride them as if the production of significant sound was their primary aim.[3] So it is not only singing that brings our world into existence and gives it meaning, but the music of technology also, together with such visual pattern-making activities as painting and sculpture, building and engineering.

Much has been written about those aspects of technology that have to do with its role in the economy, its social impacts, and its political implications. But this book is about how technology helps to endow our world with meaning—how the exploration of meanings at various levels is a source of creativity and inventiveness, and how, one might tentatively speculate, some current crises are partly the result of relying too much on technology (and too little on music and other more directly human experience) when we search for purpose and direction in life. For although technology appears to be developed for practical reasons, and evaluated in economic or military terms, its meaning and aesthetic qualities are not incidental. The visual and musical aspects of a machine may inform its design and development at every stage. They are important for the designer's motivation, and may spark his or her creativity. We judge the final result in part by how it looks and how it sounds. In the adjustment of an engine, we talk about "tuning" it, not just by analogy with musical instruments, but also because we know that when it sounds "sweet," it is likely to be running well.

Thus my argument is that, if we wish to understand what technology means to those who invent, tinker with, build, or just use its products, we must investigate how the aesthetic is intertwined with the practical; how the giving of meaning is related to building and making; and how work with tools or with hands may have some correspondence with musical experience.

It is logical, then, to start by noticing that before work was mechanized—in Europe during the industrial revolution—its pace was linked to breathing rates and other rhythms of the human body, and was often accompanied by singing, in the field or workshop, or on board ship. The sixteenth century writer, Georgius Agricola commented on the songs of miners in eastern Germany, and a modern Cornish miner has said that the acoustics of some mine galleries encourage singing.

Marion Milner records a visit to a carpet factory in Srinagar where she heard "the sound of voices chanting out the pattern, like chanting in a temple."[4] Here the singing not only set the rhythm of work but was a means of remembering complex changes in operations as an elaborate pattern was woven. Similarly, when Australians sang their landscape into existence, they were also collecting and organizing memories of landmarks and routes, and for preliterate peoples everywhere, music was an important aid to structuring thought.

Agricultural workers have commonly sung as they worked, and in West Africa, they may be accompanied by drumming. Paul Richards observed this in a Mende committee in Sierra Leone while he was measuring work rates and rice production. "In one case where I undertook measurements of the same group working on the same day with and without music, 20 per cent more work was done to drumming than without it."[5] This led to the intriguing and technologically subversive suggestion that hiring a drummer may increase an output as much as adopting the latest crop production technology.

A comparable thought occurred to a member of an Inuit (Eskimo) hunting community in the Canadian Arctic when he remarked: "Singing was just an ordinary hunting method. The Inuit used to make up lots of songs to make it easier to hunt animals . . ."[6] and to rehearse tactics. But now with guns, hunting is easier and all that is unnecessary.

Another example of how rhythm by itself (without any other aspect of music) can aid memory and improve the way a job is done comes from W. H. McNeill's reflections on army drill. After describing how the effectiveness of muskets was transformed in the seventeenth century by training soldiers to go through the complicated reloading and firing procedures in unison, he suggests that the improvement achieved arose from doing it as a rhythmic sequence. Reloading was carried out more quickly and the soldiers made fewer errors.

Weapons drill was supplemented by marching and countermarching which had a "powerful psychological effect on soldiers subjected to it." Here, McNeill reflects on his own experience of being drilled in the army, noting how the rhythms of marching submerged his own idiosyncrasies, and integrated him into the company of men with whom he drilled. There is a comparison, he thinks, with dancing, where rhythmical movements, also in unison, can make people feel closer.[7] There is similarity, too, with other tasks carried out by groups of people working together. Tolstoy described a hay meadow where grass was being mown by a line of forty-two men with scythes. Each advanced one step with every swing of the scythe, and "heard nothing but the swish of scythes behind him." The sound drew in a relative novice, until his clumsy, unpracticed strokes started to integrate with the overall rhythm, and then "it came easy to him."[8]

In working a crosscut saw, by contrast, just two people are involved, but they need to act in physical and mental unison. "If that exists, the rhythm and companionship are pleasant. One cannot talk, but unity or disunity appear through the blade of the saw."[9]

Commenting on psychological and physiological aspects of such experiences, Anthony Storr comments on how "music can order our muscular system," and for people performing repetitive tasks, whether alone or in a group, music or rhythm can nearly always enhance performance. But Storr points out that another effect of music on those who hear it is arousal, a state of readiness for action comparable to the different kinds of feeling associated with anger, fear, or sex. An electromyograph records increases in electrical activity in the leg muscles of a person listening to music, even when that person has been told to keep still. Group singing, or a shared work rhythm, or the rhythms of exercise at an aerobics class, tends to coordinate the state of arousal of a whole group, and for that reason, music was often, traditionally, a preparation for joint action: war, work, or a public ritual.[10]

Body Rhythms and Mechanical Invention

Living bodies are characterized by a variety of natural rhythms: heartbeat, breathing, walking, the daily rhythm of sleep and waking, and slower menstrual and seasonal changes. Rhythms outside the body, whether of music, machines, or nature, seem to have significance according to how they relate to body rhythms. When a puppy is separated from its mother, a slow-ticking clock placed in its basket may quiet its distress, because the sound of the clock has a similar rhythm to the mother's heartbeat. Humans, too, often find a ticking clock soothing and reassuring. A person at rest may have a heartbeat of around sixty or eighty beats per minute, which is close to the traditional pendulum clock's rhythm of one tick per second. By contrast, the quicker pace of a military march played at just over one-hundred beats per minute is more stimulating, being close to the heartbeat associated with vigorous activity.

Walking rhythms as well as heartbeat also seem especially significant. In early human history, a mother would spend much of her time walking to collect water, firewood, or food, or as part of a nomadic lifestyle, often carrying her baby on her back. One distant echo of this experience may be the way that rocking a distressed baby seems to reassure it that mother is near and all is well. One investigator devised a machine for rocking babies that operated with up-and-down and forward motion. Babies responded best when these motions were combined to resemble the effect of walking. But rocking a baby at thirty cycles per minute had no influence on its crying. Only when the speed increased to fifty cycles—nearer the normal walking rate—did most babies cry less, and at speeds over sixty cycles, crying stopped altogether. "A remarkable feature of this observation is the specificity of rate: at sixty cycles most babies stopped crying, though a few require seventy."[11]

Slower rhythms altogether are associated with breathing. Typically we take in twelve to twenty breaths per minute, and the lilt of many work songs may be related to this. However, a more complex explanation of why music and muscular activity are so closely related is that walking, swimming, and writing all involve elaborate sequences of muscular movements that are repeated regularly without our thinking about them. The coordination of this repetitive action involves more than just rhythm, and perhaps, indeed, something like a melody. So when a child learns to walk or to swim, he or she is learning a melody expressed in motion: a kinetic melody. When adults train to achieve high performance in running or swimming, they are learning a more precise version of the melody and its associated rhythms. One athlete, commenting on his poor performance in a javelin event, remarked that "the tunes were just not playing for me." He had thought hard about his performance, and he meant that almost literally. In throwing a javelin: "Finding rhythm on the run up is the most crucial thing. I have to concentrate. . . . Everything must be . . . rhythmical. It is rhythm as well as timing, and the ability to . . . create tension at the exact moment that is needed."[12]

Although the tendency since the industrial revolution has been for machines to displace skilled work, some of the most intriguing of all innovations are those that require their users to learn new body skills, and hence new muscle rhythms. Tools such as scythes, saws, and hammers must have been in this category when first introduced. So must keyboard instruments such as the harpsichord and piano. One of the most striking inventions of all is that of the bicycle, because that entailed discovering an ability to coordinate the organs of balance with the action of muscles in a novel way. The mechanical detail of the modern bicycle—its wheels of equal size and chain drive—took some time to evolve, mainly during the 1880s and 90s, but discovery of the human abilities that make cycling possible had taken longer, and arose from play with hobbyhorses going back to before 1800.

In tracing the history of hand tools, rowing boats, bicycles, hand pumps, spears and javelins, we are dealing with an aspect of the history of technology in which muscular rhythms and human abilities are all-important. In cycling, when muscular and mechanical rhythms combine well together, all is joy. When there is difficulty in getting the rhythms to fit, riding a bicycle becomes a labor. Hammers and scythes must be designed so that the length and balance of each handle is right for the appropriate working rhythm. Yet designs for such tools have varied considerably over time and among cultures, suggesting subtle differences in working rhythm.

This aspect of the history of technology contrasts sharply with the history of mechanized industry, in which we encounter inventions aimed at deskilling work and displacing muscle rhythms. These developments had an influence on the music of the period, not least because work songs became redundant and in any case were discouraged. Church music was also mechanized as organs increasingly replaced the small orchestras of village musicians that had accompanied singing in many rural churches.[13] It is worth noting that some engineers who contributed to the early development of industry were also involved in this other aspect of mechanization. Before 1766, when he became entirely involved in engineering and the development of steam engines, James Watt was in business as a "mathematical and musical instrument maker." In this capacity, he built one or two organs and invented a device for controlling air pressures within them. A century later, locomotive engineer David Joy collaborated with a Leeds organ builder in devising a new means of powering organ bellows.[14]

Long before this, however, musical rhythms were being influenced by the development of machines such as chiming clocks and carillons, and later, by a steady development of musical boxes, barrel organs and toy automata which produced musical sounds. There was a transition from the free rhythm characteristic of medieval plainsong and other early music to steady two-, three-, four-, or six-beat bars. The usual explanation is that this was "probably due to the dance,"[15] and one can certainly notice that dance tunes were used in formal music, such as early sonatas and symphonies. However, the influence of mechanical devices should also be considered, including any that produced rhythmical sounds as well as some that had musical purposes. When army drill became more rigorous in the seventeenth century, composers were increasingly asked to write music for marching, which of course required a regular beat, and later still, the rate at which soldiers drilled was sometimes checked with a metronome.

John Blacking's studies of the role of music in society contain the intriguing assertion that "the real sources of technology," like the origins of music, "are to be found in the human body"—and also in the cooperation of people in physical tasks.[16] This idea works well for the kinds of technology described above where rhythm is important, as in the use of scythes and bicycles, but does not seem applicable to electrical equipment or automatic machinery. However, some powered machines make noises that one listens to, or even enjoys, for reasons that have little to do with the human body. A person who works with a machine quickly gets to know its characteristic sounds and takes unusual notes or rhythms as a warning of malfunction. Similarly, in starting a car or controlling a motorboat, one listens to the engine, and in a cruising boat, one might aim to keep its note to a constant pitch.

Some machines have been mentioned that people like specifically because of their rhythmic action. Just as the tick of a clock can seem comforting, so could the slow pace of early (stationary) steam engines, running at less that 20 strokes per minute in many instances. But as Schafer observes, Thomas Hardy described the faster, "inexorable" rhythm of a threshing machine as being able "to thrill to the very marrow all who were near." And D. H. Lawrence found the rhythmic noises of a mine winding engine "startling at first, but afterwards a narcotic to the brain."[17]

There is a distinction here between relatively slow rhythms (20 to 130 beats per minute) which can vary in their effect from soothing to stimulating, and the faster vibrations of a throbbing engine (usually more than 150 beats per minute). Harley-Davidson motorcycles have old-fashioned, slow-revving engines that "come on song" (as enthusiasts say) when they are opened up, to run with a slow, even beat and "a wonderfully resonant exhaust sound." This seems to encourage a characteristically relaxed camaraderie among owners, who manifestly enjoy modulating the engine note, and who ride around in a relatively leisurely way, quite different from the speeds at which high-revving Japanese bikes tend to be ridden. Some diesel engines produce a slow, satisfying throb—notably the Lister engines used on English canal boats—and some car engines are said to "purr."[18]

With vibrations of higher frequency (more than, 1,500 beats per minute or 25 Hz), we are no longer aware of either rhythm or throb, but instead we hear a buzz, hum, or continuous note. Electrical equipment tends to make this sort of noise. Sometimes we learn to ignore it, but otherwise it is monotonous, and can have a depressing, headache-producing effect.

Turning from the rhythmic quality to the loudness of noises associated with technology, it sometimes seems that noise and power go hand in hand, and that during the industrial revolution, factories were permitted to produce a great deal of noise without anyone complaining because the power of their owners was largely accepted. The long historical tradition of using trumpets and drums in warfare is also significant as an expression of power. These instruments were a means of evoking fear by making intimidating noises at the start of a battle. Part of the attraction of guns when they were first invented was that although they were too inaccurate to damage the enemy in any reliable way, they added greatly to the fear-inducing noises an army could make. Several authors have noted the importance of this, and some have even claimed that "if cannon had been silent, they would never have been used in warfare."[19]

Today, advertisements for cars have sometimes presented noise as an advantage, particularly when the roar of a powerful engine is stressed. In the Japanese automobile industry, the noises made by cars are sometimes deliberately engineered to be attractive to potential purchasers. The "aggressive" kinds of noise are reduced in cars intended for sale in Japan, but vehicles made for export are "tuned" to meet tastes elsewhere.

Music making has both reflected and reacted against these developments. Large orchestras became more common during the nineteenth century, partly, it is said, in reaction to the growth of industrial noise. Pianos evolved from relatively quiet, lightly built instruments and became rather massive machines. More recently, the development of rock music may be quite strongly indicative of both positive and negative responses to industrial and urban noise, and the "ferocious acoustical environment" of modern life.

Music and Mathematics

The foregoing comments on marching, muscular activity, and work songs may suggest that music has a physical rather than an intellectual significance. Blacking's studies point to "a rhythmic stirring of the body" as the beginning of music, and describe how it also helps people to feel a certain solidarity with one another in work, dance and ritual. "Many, if not all, of music's essential processes can be found in the constitution of the human body and in patterns of interaction of human bodies in society."[20]

Yet there is a long-held belief that music has meaning in terms of mind as well as body, and that it is closely allied with that most intellectual of pursuits, mathematics. This belief goes back to Pythagoras and his associates in the sixth century B.C., who are said to have studied how the notes produced by stringed instruments vary with the length of a string that is free to vibrate. Halve the length of the string and the note changes by an octave. So the ratio 2:1 corresponds to an octave, and the Pythagoreans found that other simple ratios (for example, 3:2) were also related to intervals in the musical scale (in this case, the fifth).

It is significant that the Pythagoreans envisaged a link between their mathematical view of music and muscular and other functions of the human body as well, for they regarded the body as a kind of musical instrument in which each string must have the right tension. But, it may seem to us more strange that they looked for the simple ratios associated with music in the structure of the universe, and expected to find them reflected in the orbits of the planets. Music gives us a strong feeling that life has meaning. The starry sky suggests meanings of other kinds, and it was tempting to think they were related.

We can trace such convictions from Pythagoras to Plato, and to the great geometers of the ancient world, and we find them reemerging in the work of medieval and Renaissance thinkers. Within this tradition, belief in a critical connection between mathematics and music came to a head around 1600 in the work of two key figures in the history of scientific thought, Galileo and Kepler.

Galileo was a critic of number mysticism in the theory of music. Kepler, however, remained committed to the old view of a link between simple ratios and musical harmony. A culminating point in his career was publication in 1618 of a book he wrote on "the harmony of the world." This work began with an interpretation of the simple ratios of Pythagorean music theory. Then, after a long digression, the argument focused on the motions of the planets in orbit around the sun. Kepler had already shown that the orbits of the planets were elliptical, and that a planet's speed of motion varied as it went round the ellipse. These were discoveries of great importance, but Kepler interpreted them by looking for simple ratios within the new structure of the solar system he was describing. And he claimed that the ratios associated with musical harmony seemed to be reflected in the motions of the planets through the sky: "the heavenly motions are . . . a continuous song for several voices." This poetic depiction is dismissed by Arthur Koestler as a "luxuriant growth of fantasy."[21]

Koestler expresses surprise that important scientific discoveries emerged alongside such notions. But ideas about the music of the spheres had gone very deep into European culture. Hildegard of Bingen, a medieval musician and philosopher, said that "Music is a sense of heaven. All Creation is filled with music, resonating through the spheres."[22] Emotionally, nature can still have this meaning for us, but we nowadays separate the emotional and the scientific in ways that Kepler did not. Even so, Kepler lived at a time when people were detaching themselves from such enchantments, especially people who thought like Galileo.

Galileo argued that instead of making simplistic assumptions about numbers and ratios, we should study what was actually happening when strings on an instrument were vibrating. He analyzed vibration in much the same way as he discussed swinging pendulums.[23] On that subject, he demonstrated a fixed time for a particular pendulum to complete its swing, which could be used to regulate a clock so that it would keep time accurately. Indeed, the first successful pendulum clocks were made in the 1650s, after his death.

As a youth, Galileo had helped his father, the musician Vincentio Galilei, with work on the tuning of strings. However, Vincentio's views, like Kepler's, were oriented to the music of ancient times,[24] and he was critical of modernizers who wished to tune instruments with anything other than Pythagorean intervals as defined by simple ratios. Apart from the octave (with its 2:1 ratio) and the fifth (3:2), he quoted figures for major and minor thirds (5:4 and 6:5), for the major sixth (5:3), and even for the semitone (16:15 or 25:24).

A modern commentator has remarked that measurement entered music in the seventeenth century,[25] and we can see the beginnings of that process, perhaps, in Galileo's work on the pendulum, for that made the metronome possible by 1696. Watches with balance wheels were being developed at about the same date, and in 1724 it was said of one piece of music that the crotchets should be "counted as fast as the regular motion of a watch." However, only toward the end of the eighteenth century did composers begin to recommend metronome settings (expressed as crotchets per minute) on the scores of their music. So now music might sometimes be played—and armies might be drilled—by the precise rhythms of a machine. It was about this time also that people employed in the early factories were having to work to the rhythms of other sorts of machines, instead of to their own singing.

Much of this musical mathematics is a matter of convention and has no deep significance. Although the octave is common to most types of music worldwide and has a basis in the physics and mathematics of vibration, ways of subdividing the octave vary enormously. Many cultures do not use the Pythagorean intervals in their scales, and there is no reason to think that their music is worse for that. So if claims are to be made about a fundamental affinity between music and mathematics, they must refer to something deeper than scales and metronome markings.

Some of what might be said to account for this affinity is first, that music and mathematics both deal with pattern, expressed in abstract terms, and second, that mathematics and music both engage their practitioners in strong feelings about beauty or elegance. Of course, music also deals with other emotions and physical reactions, not just the kinds of feeling experienced by mathematicians. It is also much more accessible to those who are not practitioners.

Third, though, music and mathematics both involve organization. "Players" in both fields rearrange and organize the patterns they deal with to create or discover new kinds of order—beautiful, or at least meaningful, order.

These points are nicely illustrated in Douglas Hofstadter's study of parallels among mathematics, music, and visual pattern making, the latter disciplines represented especially by J. S. Bach and M. C. Escher. Some parts of Hofstadter's arguments are illustrated by dialogues in the style of Lewis Carroll (Charles Dodgson), which involve a character named Crab, a musician. On one occasion, after Crab has given a flute recital, Achilles happens to see his sheet music and is startled to observe mathematical symbols where he expected to see crotchets and quavers. Naturally, he enquires about this, asking Crab: "Are you sure this is musical notation?" But Crab insists that it must be music because he has just played from it. Indeed, not being a mathematician, he could not have understood it had it really been mathematics.[26]

The point is thus made that mathematicians and musicians deal with patterns that, on a sufficiently formal level, may be of similar kind.

Moreover, Achilles also raises the point about mathematical elegance. He suggests that if a proven mathematical truth were to be written in Crab's notation so that it could be read as music, then it would surely be beautiful music. But if a false mathematical theorem were read as music, that would probably be jarring and discordant.

Anthony Storr, for whom music is the "fundamental human organizing activity," makes a similar point about correspondences between mathematical truth and musical beauty, and then asks whether patterns in music, and in mathematics, are "human invention, or are they discoveries of some pre-existing order?"[27]

Mathematicians tend to feel that many of their equations reflect the ordered structure of the universe. Is this feeling justified, and are mathematical relationships inherent in nature, waiting to be discovered? Or do mathematicians simply invent their patterns and impose them on nature? Kepler's work on the orbits of the planets seems to be very much of this second kind. He found ratios characteristic of musical harmony because he felt that such ratios must be there, and he kept trying different numbers until he got a fit between planetary motions and Pythagorean ratios. In other work, he fitted geometrical shapes within the orbits of the planets, painstakingly adjusting them to match his assumptions about mathematical regularity. His insights were poetic but his mathematics seems contrived. Yet his elliptical planetary orbits are accepted as true and were reached by the same methods. Kepler's results were clearly invented by him and imposed on what he observed. Are mathematical relationships and scientific theories always invented like this rather than being truths that are discovered?

All we can really say is that some mathematical regularities we think we see in the universe are not really there. Others, on the other hand, fit so neatly and work so well that their reality is hard for even the most confirmed skeptic to resist.

Two mathematicians who have thought hard about this issue comment that the different kinds of mathematics we use to describe the universe are "all human invention," but in a special sense.[28] Rhythms and mathematical relationships are patterns that we find within ourselves through the use of our brains. But each human brain is located in a body and is part of the natural world it seeks to understand. There are correspondences between what we find within and what we observe outside ourselves because brains are part of the world we observe. They have also been selected (in the Darwinian sense) for an effective fit with reality. It should not be surprising, then, that the mathematical patterns that occur to us when we are using our brains are sometimes congruent with patterns that exist on a large scale in the universe. Provided that we can select the right patterns, it is understandable that we can find mathematical laws of nature that seem to work.

Could it be, then, that musical phrases and rhythms are part of the same scenario? Could it be that all pattern-making in time, whether produced by nature or technology or art, relates to the same underlying dance? One view of science is that physicists and others working in the most "fundamental" areas of enquiry do not really explain the world: "they are only dancing with it."[29]

What most authorities seem to agree is that music reflects in some way the order—the organization—that is necessary for the human nervous system to function. Hofstadter, with his interest in artificial intelligence, suggests that music is related to the "software" with which our bodies and brains are "programmed.." That is an instructive comparison even though brains are not really much like computers, and it becomes an even better parallel when Hofstadter turns it around and comments that J. S. Bach's *Musical Offering* "reminds me . . . of the beautiful, many-voiced fugue of the human mind." One may feel that something similar is intended by Anthony Storr when he quotes Hegel's comment that music is an analogue of the "inner life," and Michael Tippet's view that music embodies the "otherwise unperceived, unsavoured inner flow of life."[30]

Another point that may throw light on whether music deals with something fundamental in nature, or whether it is merely what human beings do to entertain themselves, may be indicated by the facts that, first, human music can have a soothing effect on animals of several species (e.g., cows and even lions); second, that some mammals, especially whales and gibbons, communicate by complex vocalizations that have marked affinities with music; and most significantly of all, that birdsong shows many formal similarities with human compositions. Research on this last subject[31] shows that if the very rapid sequences of notes produced by a skylark or blackbird are recorded, and then replayed more slowly, some of the same formal patterns as occur in human music can be recognized. Observations made using a different technique, the sound spectrograph, had earlier shown that late in the breeding season, a blackbird's song becomes more closely organized in a manner "identical to that which we find in our own music."[32]

Human beings and birds are not closely related in terms of their evolutionary history, but similar rhythmic patterns may perhaps characterize the nervous system of all higher animals—which is why some find music soothing or relaxing even though they do not produce similar sounds themselves. Birds differ, though, in that many of their body rhythms (including heartbeat) run much faster than in humans, and the rapid wing motion of some shows that the rhythms of muscle coordination must also be faster. It is not surprising, then, that birdsong has to be slowed down before its congruence with human music can be appreciated. Also, birdsong includes smaller gradations in pitch—"microtones" rather than semitones. Some human music uses intervals smaller than a semitone, of course, but avian microtones are much finer intervals that cannot always be distinguished by the human ear.

Some say that birdsong is not music, because it conforms to predetermined, "instinctive" patterns without free composition. This may not be entirely true, because some species, larks and thrushes especially, do seem to improvise elaborate sequences of notes. And although many human composers have quoted phrases from birdsong in a casual way, Olivier Messiaen used avian forms extensively and sensitively, as if he recognized real musical quality. So when the question is asked whether human music corresponds to anything outside the human mind, one answer must be, yes, there are close parallels with avian (and some mammalian) vocalizations. One must therefore wonder about rhythmic and melodic pattern as a common feature of many animal nervous systems.

Two Kinds of Meaning

Although I have quoted mathematicians, biologists, historians, musicians, anthropologists, and writers of fiction in this chapter, really only two kinds of discourse about music are represented here. On the one hand there are explanations of what music does and how it interacts with other human activities. On the other hand, though, there are descriptions of what music feels like. Hildegard of Bingen, the medieval philosopher and composer quoted earlier, described her musical self as "a feather on the breath of God," and somebody from a quite different culture has exclaimed: "All my being is song. I sing as I draw breath."[33] A modern, Western view is that: "Music is an irreplaceable, undeserved, transcendental blessing."[34] These exclamations can be compared with descriptions of what it feels like to work in various branches of technology, or just to use tools, such as the scythe or the cross cut saw mentioned earlier, or to ride a bicycle. In many instances there is a close parallel with musical experience when muscular and mechanical rhythms combine well together. Different but complementary is the experience of order and pattern in mathematics, and in complex engineering structures, as perhaps also in a fugue or symphony.

One of the most moving descriptions of what music feels like was written by the neuropsychologist Oliver Sacks after a serious accident to his left leg that ruptured a tendon. Two weeks after surgery, the leg remained an immobile, inert object in a plaster cast. When physiotherapists tried to help him walk, he could do so only by thinking out each step before he took it.

But a friend had given him a tape of Mendelssohn's violin concerto, to which he listened in bed, and from the moment it started, something happened: "The music seemed passionately, wonderfully, quiveringly alive . . . as if the animating and creative principle of the whole world was revealed, that life itself was music."[35]

A few days later, Sacks was with the physiotherapists, struggling to walk in a clumsy, unstable, robotic way, when suddenly, a tune began to run through his mind: "Mendelssohn, *fortissimo!* . . . And, as suddenly . . . I found myself walking, easily, with the music. . . ." As the natural, unconscious rhythms of walking came back, it was like "a return of my own personal melody, which was somehow elicited by, and attuned to, the Mendelssohnian melody." What seemed to appear with the music was organization and a center, coordinating the different functions of muscles in his leg and making them work together as a whole. Moreover, the experience restored a lost sense of the meaning of his own self. What announced itself in music was "vital feeling and action," transcending the physical.[36]

If Sacks had been a "better" scientist, more adequately equipped with inhibitions designed to make him objective and impersonal, he would have written about this experience far more cautiously and in quite different terms. For example, he might have explained exactly how rhythms may be involved in coordinating muscle action, quoting theorists who have constructed mechanical models on paper that describe the nervous system as a network of "linked oscillators." Walking can then be analyzed in terms of several rhythms controlled by a central "pattern generator" somewhere in the central nervous system (not necessarily in the brain itself). The theorists acknowledge that this is all hypothetical, but claim that something like it must occur.[37]

Had he discussed these ideas, Sacks could have described his own experience far more clinically, pointing out that the Mendelssohn concerto may just have given a nudge to his own pattern generator. This more scientific account, conjectural and labored though it may be, could then have led to discussion of the growing field of music therapy, in which many patients who suffer from defective muscular coordination, or problems of motor control, can be helped by the "dynamism of music."[38]

All this is valuable, but surely we would have missed a lot if Sacks had written in these terms without telling us what it felt like for his recovery to be aided by music. And in telling us that, he has given an account of what may properly be called his *existential experience* of recovery, by contrast with the scientific account he could have given had he stepped outside the experience and described it from the viewpoint of a professional neuropsychologist.

That brings us to the critical point in this chapter. No aspect of human life, be it music, medicine, or technology, can be adequately discussed if we are always restricted to a scientific mode of discourse. If we wish to discuss a human activity, there are times, as Sacks beautifully demonstrates, when there is more insight to be gained from knowing what something feels like—knowing what its existential meaning is—than from knowing how it works and measuring it.

But discussion of technology from this point of view is very uncommon. Samuel Florman's memorable book *The Existential Pleasures of Engineering*[39] is an example I follow in looking to literary and autobiographical accounts of technological experience for descriptions of what it feels like. But I am suggesting in this chapter that musical expression is also relevant, particularly for conveying a sense of what the action and the order in technology feel like. Work rhythms, a feeling for structure, and the sense of controlled power conveyed by an organ prelude or, in a different mode, by rock music, all say something about the experience that informs much technology.

A Musical View of Nature

One way of explaining what this book is about, then, is to say that it describes what technology *feels* like to its practitioners (Chapters 2 and 3), to consumers of its products (Chapter 4), and in relation to the environment (Chapter 5). The book also says something about how technology feels in the context of relations among people (Chapter 7), and in manifestations of violence against people (Chapter 8). Describing such feelings could serve a useful purpose, I believe, in making us more self-aware and more conscious of why we respond as we do to new technology or to the impact of technology on the environment. But it might also seem self-indulgent to devote a whole book to experience that many would dismiss as "merely subjective."

There is rather more at issue, however, because many people are aware of the discontinuity that we have just noted in the work of Oliver Sacks. Describe the world in scientific terms—in the language of neuropsychology where the leg injury was concerned—and one is able to look for explanations. Here the word "explanation" means "finding the cause" of what has gone wrong. In modern culture, "explanation" also means describing what happens by reference to a mechanical model, such as the model involving linked oscillators that other scientists (not Sacks) have used to describe the process of walking, and the models of mechanical vibration that Galileo used to explain why some musical sounds are pleasant, some discordant, and some combine "the impression of a gentle kiss and of a bite."[40] That makes music a matter of the mechanics of physical sensation.

Although this book is mainly about technology, we have to recognize that our habits of thinking about most technical subjects have been shaped by science. It is worth noting, then, that many of the viewpoints we need to discuss came to the fore during the "scientific revolution," which was a more gradual development than its name implies, overlapping both ends of Galileo's life span: 1564–1642. It was a time, though, when traditional ideas about nature and how it should be studied were reoriented and reformulated in quite radical ways, and there are many strands in the complex changes in outlook, of which three have so far been alluded to.

First is the stress on objectivity, which means that when we discuss a leg injury, or analyze music, or study any natural phenomonon whatever, we strive to separate our personal experience from the outward, observable manifestations, and we study the latter. This is a valuable intellectual maneuver that makes it clear what to record about a phenomenon, and what ideas about it might be tested by experiment. But we need to be aware of the loss of insight that may result from habitually sifting all experience this way. Second, there is the view that to explain something, we need to find its cause, the implied assumption being that everything must have a cause. Third, there is a tendency to think that the most likely causal explanations of events in nature are to be found by making comparisons with machines. The generation following Galileo spoke of science as a "mechanical philosophy" and discussed the human heart as a pump, arms and muscles as levers and springs, and insects as clockwork automata.[41]

Although science became strikingly more successful by adopting these ideas, it did so at the cost of marginalizing a range of experiences (notably of music) that clearly had meaning for many people. One might almost say that science achieved its successes by disregarding the humanly significant. However, the aspect of this that philosophers are most wont to worry about is not music, but the experience people have of being free to make choices and to pursue their own purposes. According to the outlook in which everything is seen in terms of cause and effect, a person's actions are not chosen but caused. How, then, does the experience of having a purpose, and of making choices, arise? Similarly, how does the experience arise of music representing organization and purpose, and the "inner flow of life?" There is a general question, then. Why do we feel that music is purposive, that work should be purposive, and that choosing is purposive when objective scientific analysis does not consider purpose at all, but offers alternative explanations of music and human behavior?

It has been said of Bach that, "in his music one is seldom totally divorced from a motor rhythm . . . regular, reiterated, non-developing." But soaring above it, "independent polyphonies sing and wing, often transcending, sometimes even contradicting the beat."[42] In such music, then, experience of regular, predictable, caused events coexists with perceptions of spontaneous, exuberant change, and perhaps of purpose.

During the time of the scientific revolution, there were arguments about whether compass needles (for example) pointed north because they had some kind of animate purpose in doing so, or whether they were moved by some unseen cause. There were also questions whether metal ores grew in the earth like living plant forms. Such ideas were being discussed around 1600, notably by William Gilbert (a physician interested in the magnetic compass), and by metallurgists influenced by alchemy.[43]

It was a clear gain for science to say that rather than treat everything as if it were living and moved by animate purposes, investigators would do better to treat living things as if they were merely mechanical. That meant regarding plants and animals as having no potential for purposive action, but only a capability of responding to external stimuli. These stimuli could then be expected to "cause" behavioral effects according to fixed laws of nature.

At this point the claims quoted earlier about music reflecting something fundamental in nature seem especially relevant, because although "cause" seems to be a more powerful explanatory concept than "purpose," it does not capture all our perceptions. There is in nature an element of the spontaneous and the purposive such as we also experience in music.

The paradox is that whenever people have tried to describe nature in ways that would allow it to have a defined purpose, they have ended up on a false trail. This applies to all pictures of the physical universe that had God intervening actively to keep the planets in their orbits and in other respects to keep it functioning in accord with his purposes. It also applies to vitalism in biology, and to all ideas suggesting that the evolution of living things has a defined purpose or goal. But just as music can be powerfully purposive in its energy, direction, and organization *without* having a defined goal, so also seems nature.

In this way, it is possible to accept a scientific worldview and appreciate nature as neither animate nor guided by divine purpose, while at the same time seeing the natural world as having spontaneity, direction and an "inner flow of life" in the same sense as music has these qualities, but machines do not. It is possible, then, to suggest a musical model differing significantly from the mechanical models that have predominated since the scientific revolution. It would still, of course, be a model rejecting the animistic view of the world that the scientific revolution displaced, but in a less reductionist way.

This idea of a third way between primitive animism on the one hand and mechanical philosophy on the other has emerged here from a rather impressionistic discussion of music. A comparable but far more rigorous account of purposiveness in nature has been developed by Howard Rosenbrock, an engineer who has written extensively on the philosophy of technology. His experience of designing control systems to allow machines and industrial processes to run automatically is that in this area of technology, human purposes and goals have to be translated from the purposive conceptual framework of ordinary human affairs to the mechanical, cause-and-effect worldview presupposed by modern engineering disciplines. Within these disciplines, Rosenbrock says, "All of nature is seen as a machine without purpose, though each person makes a lenient exception in his own interest."[44] That is, although we are part of nature, we except ourselves from the mechanical view of nature when thinking about our own goals in life and the choices that confront us.

It seems fundamentally unsatisfactory to be bound to mechanistic thinking when one is so aware of the experience of purpose in one's own life.[45] What lies at the root of this contradiction? Is experience mere delusion, or is nature more subtle than we think?

Rosenbrock suggests that fully to capture the reality of nature would ultimately require a theory more "abstract and tenuous" than we could handle.[46] To make any progress, we have to clothe the abstract in language, or mathematics, or sometimes in visual images. We often have to use words that refer to a cause-and-effect concept, but if that leaves us feeling uncomfortable at times, it is because the reality it deals with is more subtle than it can quite describe.

Rosenbrock does not mention music directly, but it seems reasonable to suggest that music is another way of "clothing" perceptions that would become too tenuous if we tried to use them directly. As for technology, before engineers had mathematical training and could use data from books, journals, or computerized databases, they were craft workers, dealing directly with timber, iron, or stone, and clothing their perceptions in material form without necessarily translating them into words or numbers at all, as we shall see in the next chapter.

The worldview that I presuppose in this book, then, portrays technology as part of a more subtle nature than causal mechanical models allow. It is a worldview that makes sense of the point often remarked on, that scientists and engineers often seem especially knowledgeable and aware of music, as if there were something in music to aid scientific development. And it is a worldview in which human relationships and human purposes may have a closer connection with technological progress than sometimes seems possible.

2

Visual Thinking

Visualization in Science and Engineering

The creativity of scientists and inventors has long been debated by psychologists, and among classic studies that are still frequently quoted[1] is one by Anne Roe. She describes interviews with scientists active in the United States around 1950 who were then regarded as eminent. Some of these people were surprising to Roe because they seemed to be using thought processes "quite foreign to my own." Some, indeed, appeared unable to verbalize freely and depended on visual imagery more than words.[2]

Investigating this further, Roe eventually classified the thinking processes of about sixty scientists (physicists and biologists) according to whether they mainly used visual or verbal imagery. As one might expect, most combined different modes, but the experimentalists in the sample mentioned visual imagery most often, whereas theoretical physicists more frequently employed verbal imagery, their thinking being rather like "talking to oneself."[3]

Of those whose thought processes were predominantly visual, not everybody used visual images of the same kind, however. Some scientists referred to "concrete," three-dimensional images; others spoke of more "diagrammatic" images; and some merely visualized formulae or symbols. A few scientists also mentioned that some of their thinking did not have either visual or verbal content. They experienced something going on in their minds that led to results but involved no images, and Roe described this as "imageless thinking." She did not make comparison with music, but her description recalls the feeling one can have when listening or playing that a wordless ordering and structuring process is under way.

A few years before Anne Roe began her work, a less rigorous enquiry into the working habits of mathematicians included Einstein within its scope, and his reply to the questions asked is often quoted. He said that "words . . . as they are written or spoken, do not seem to play any role in my mechanism of thought." Instead, the elements in his thinking were "more or less clear images," which were of "visual and some of muscular type". Having worked out his ideas in these terms, Einstein found that conventional words or mathematical signs had to be "sought for laboriously . . . in a secondary stage."[4] The instigator of this study, Jacques Hadamard, came to the general conclusion that among the mathematicians he had contacted, all of them born or resident in America, most tended to avoid use of words in their thinking, or even algebraic and similar signs. Rather, they worked with images that were "most frequently visual, but they may be of another kind, for instance, kinetic."[5]

Once again, we are given an impression of the importance of visual imagery for thinking in a scientific field, but with various other less well-defined thought processes indicated by such words as "muscular," "kinetic," or elsewhere in Roe's work, "kinesthetic." Engineers, inventors, and other technologists have not attracted the attention of psychologists to anything like the same extent,[6] but some excellent historical studies help fill the gap. From these, it is quite clear that (as might be expected) nineteenth- and early-twentieth-century engineers habitually used images of the more concrete kind, visualizing the artifacts they designed in an immediate, three-dimensional way.

One historian, Brooke Hindle, has studied early steamboat development, from about 1780, and observes that nearly all the inventors and engineers in this field had previously been active in art, architecture, or mapmaking, and employed skills that they learned in these visual fields.[7] Eugene Ferguson, another historian, has examined the work of engineers who spoke explicitly about how they could visualize the machines they designed in their "mind's eye." Thus Walter P. Chrysler, founder of the automobile firm of the same name, could recall building a model by copying the image that existed "within my mind so real, so complete" that he had no need to work from drawings.[8]

Testimony of these various kinds is immensely useful for the purposes of this book because it tells us a good deal about what it feels like to be an engineer, mathematician, or physicist. However, some historians of science and technology dislike the psychological approach, and dismiss descriptions of what is seen in the mind's eye as too subjective to be of interest. According to their more positivist view, it is only justifiable to discuss visual thinking if there is some tangible evidence, for example, in the form of drawings made as part of the thinking process.

Such evidence is available for the development of the first telephones during the 1870s. It consists of numerous sketches by Thomas Edison and Alexander Graham Bell that Bernard Carlson and his colleagues have analyzed.[9] Their studies make it clear that the two inventors employed drawings for both social and cognitive purposes, that is, for communication with colleagues on the one hand, and for thinking on the other. A block in his thinking, for example, could prompt Edison to make several quick drawings representing a range of possible solutions—ten variant telephones on three sheets in one instance. Or he could use sketches for doing "thought experiments" that allowed him to test an idea without building the device. Bell had less skill in visualization and made fewer drawings, and so had to test almost every idea physically.

Analysis of sketches and drawings can in this way illuminate more clearly what kind of thinking may be carried out in visual terms. But in this chapter, we not only pay attention to the functions performed by visual thinking, but we also consider what this style of thinking is like as an experience. From that point of view it is still relevant to note what engineers say about the visual imagery that is part of their thinking.

Research of the kind Anne Roe pursued may also seem unfashionable because advances in cognitive neuropsychology have provided a more precise and detailed way of understanding some of the relevant thought processes. Although Roe noted that there seemed to be at least two different kinds of visual imagery, which she called "concrete" and "diagrammatic," it was not at all clear why this distinction was significant. Now, however, diagrammatic reasoning is better understood, and we know that separate parts of the brain deal with different kinds of visual material. This may be the case particularly with the spatial properties of an image, as distinct from properties such as color and texture.[10] It becomes clearer, then, why scientists who can think effectively in terms of spatial or diagrammatic images do not necessarily make use of concrete images related to the color and physical appearance of what is seen. Yet the earlier research continues to be of interest for its descriptions of experience of the different kinds of visual thinking.

Drawings and Conceptual Models

Much of what is said about thought processes in science and invention refers to puzzle-solving or problem resolution of a kind that depends on recognizing a structure or pattern in a quasi-visual way. This could be part of what the mathematicians quoted earlier may have been talking about. But it is precisely this kind of issue for which it is hard to find data adequate for a rigorous study.

Historical evidence for visual thinking more often refers to concrete, pictorial images related to design, invention, or conceptual models in science. It is then easy to see how artistic and technical skills are related. According to Brooke Hindle, a sense of spatial relationships and "an eye for detail" are just what one finds among the artist-engineers active in the early development of steamboats.

Hindle also quotes Samuel B. Morse as an artist-inventor, describing how he got the idea for an electric telegraph signaling with dots and dashes during conversations he had on a ship bringing him back from a visit to Europe in 1832. He visualized and sketched a complete telegraph system while still on the ship, but without specifying materials or dimensions.[11] Not until 1835, when he became professor of painting and sculpture at New York University, did he acquire colleagues with the technical expertise necessary to make the telegraph work.

Many nineteenth-century scientists as well as inventors kept sketchbooks, or made more formal drawings, or depended heavily on others to make drawings for them. There should be nothing surprising about this, because it seems reasonable to think that much science would be impossible without visualization. As sociologists have put it, in their inimitable way, "depictions are constitutive of scientific production."[12] The point is made, indeed, that scientific disciplines have sometimes been inaugurated by the invention of visual languages. Certainly, different phases in the development of sciences such as chemistry, biology, or geology show evidence of different kinds of visual awareness.

For example, a scholar who has studied the botanical work of Carl Linnaeus suggests that not only was "a strong visual memory" essential to Linnaeus's work and an ability to handle visual images and "concrete mental pictures," but that for Linnaeus, new ideas stemmed from visual metaphor notably with regard to sexual functions in plants.[13] In physics, "concrete images" had to be invented to model the behaviour of atoms or light rays, and Michael Faraday had to refine depictions of "lines of force" around magnets or electric wires.

At first, Faraday had talked about "magnetic curves," but then came to see the whole of space patterned with electric and magnetic lines of force. His visual style was of considerable importance for his way of working and was accompanied by a range of other visual interests. Thus he collected portrait prints, attempted to learn perspective drawing, and took an active interest in new technologies for reproducing visual images, particularly lithography and photography. Faraday was unusual among contemporaries and colleagues in making very little use of mathematics, not because he lacked mathematical ability, but because he could think rigorously about changes in a physical system by "reconfiguring each state in his imagination." He also made physical models in his laboratory, including some that suggested the principle of the dynamo and electric motor. Hermann von Helmholtz later commented that it was "in the highest degree remarkable" that Faraday was able to arrive at so many discoveries in this way, which would normally require mathematical deduction. It was "a kind of intuition."[14]

Similarly, nineteenth-century chemistry became more powerful in solving problems and yielding fresh insight as its visual vocabulary developed. Following John Dalton's work, published in 1808, it became possible to envisage each element as having its own distinctive type of atom. From then on, people could consider the possibility of making visual models of molecules by writing formulas, or using schematic diagrams, or making ball-and-stick models. Listening to modern chemists talking about the philosophy of their subject, I have been left with a strong impression that, even when they do not acknowledge the role of visual language, it may often be what decides whether a new idea can be entertained. For example, recent biochemical theories describing how cells turn food into energy have depended on visualizing what happens as electrons and protons pass through the walls of cells. Discussion may be entirely verbal, yet it is full of visual imagery referring to loops, half loops and flow pathways.[15] Ability to visualize complex molecules in three dimensions is also recognized as important for understanding reactions and designing new drugs, today using computer graphics to assist the visualization process.

Geology is another science that began to take shape only when its characteristic visual language—its diagrammatic sections, depictions of rock strata, and mapping techniques—began to develop just before 1800. Previously, a great deal of information had accumulated, along with collections of fossils, but the formulation of effective generalizations was problematic. Technical drawing made a contribution when some of the earliest geological sections were drawn by surveyors and engineers, notably William Smith and John Farey. But they tended to use ruled lines to represent strata that were not nearly so regular as this engineering convention implied.[16]

In the next generation, Charles Lyell expressed frustration about his lack of ability as a draftsman, and the crude drawings in some of his early work may have limited his ability to visualize complex rock formations. After he married, though, his wife, Mary Horner, did a great deal of drawing for him (as did other geologists for their husbands).[17] This work was not just a matter of recording things seen, or of book illustration, but further developed a visual language so that geological problems could be more adequately thought through.

In technology, visual thinking and appropriate visual languages were even more important, because many practical problems "cannot be reduced to unambiguous verbal description" and must unavoidably be thought about by means of "a visual, nonverbal process." Mathematical analysis of a bridge, a dam, a machine, or a transportation system is always based on simplification, and if an engineer is to make good judgments about the design and construction of such artifacts, his or her thinking must include the commonsense check on realism that visualization can provide. Eugene Ferguson, indeed, argues that the mind's eye offers an opportunity to review "the contents of a visual memory," and the chance to check a new design against past experience.[18]

Contrasts with Verbal Thinking

One critic argues that what Ferguson characterizes as visual thinking "is not thinking in the sense of conceptualizing or relating concepts; it is thinking as picturing."[19] One reason for this kind of reaction is the assumption that our deepest insights can always be expressed in words or mathematical equations. So "thinking as picturing" is not usually acknowledged to be real thinking at all. It was perhaps better understood in the past, when several engineers referred to drawing as providing the "alphabet" or "language" necessary for communicating their ideas. Ferguson notes that Swedish engineer Christopher Polhem constructed "an alphabet of wooden models," and Robert Fulton, the steamboat pioneer, also thought of levers, screws, wedges, and the like as an alphabet. Later American inventors picked up visual images from patent specifications and technical literature, and these evolved to become a "visual or spatial alphabet consisting of mechanical, electrical-circuit, and chemical-process symbols."[20]

Among the many nineteenth-century engineers who spoke of an engineering alphabet based particularly on drawing, James Nasmyth merits closer attention because he was so aware of his methods and so clear in explaining them. He recalled how his visual skills developed under the guidance of his father, his sisters, and the engineers with whom he worked. His father, landscape painter Alexander Nasmyth, was very emphatic in describing drawing as a "graphic language" that should be part of everybody's education because of its importance in cultivating observation and memory. His six sisters, some of them also distinguished artists,[21] practiced what their father preached by running drawing classes in Edinburgh. One of their pupils, later well known as a mathematician, was Mary Somerville (then Mary Fairfax). The drawing classes encouraged her mathematical studies through what she heard there about geometry as the foundation of astronomy and other sciences.[22]

Nasmyth the engineer learned much from his remarkable family, but also described how he learned from the "graphic" thinking of Joshua Field, an engineer who sketched all the time when talking, using drawings to reinforce words. Nasmyth was himself skilled at the workbench, and there found that "the eye and the fingers—the bare fingers—are the . . . chief sources of trustworthy knowledge." And he noticed that on occasion, images of machines would come into his mind fully formed, without the intervention of words, as if a picture had been handed to him. Thus the steam hammer for which he became famous had been "in my mind's eye long before I saw it in action," and before he described or drew it.[23]

Not only is the ability to visualize, or to develop ideas by drawing, sometimes referred to in terms of visual language and thought, but it is also occasionally characterized as "nonverbal" thinking.[24] To some people that sounds like a contradiction in terms, because of a widespread impression that words are essential to all thought. Drawings and models may clarify what we say in words, they argue, but then add that graphic language can never replace words.

This is an issue I have been able to understand best through autobiographical reflection. Working in a physics laboratory during the 1950s, I developed ideas mainly when drawing graphs. Later in life, I learned to use scale drawings in historical studies of buildings and industrial sites, and again found that drawing was an activity that prompted thought about the interpretation of my material. In some moods, indeed, it feels as if nothing can be properly understood until it has been drawn, and even a diagrammatic representation of concepts or definitions can seem important.[25]

Most of the ideas that arise from such work are expressed in words, but they would not usually have occurred to me at all if I had not been drawing. So the critics might be right in saying that visual imagery is usually just a crutch for verbal thinking, and that it is not really "visual thinking" in any independent and distinctive sense. But there are occasions when visual reaction may bypass verbal thought and expression entirely. One instance occurred on a day when I had been making drawings in pen and pencil, but then somebody lent me a box of oil pastels. After working in black and white, the vividness of the pastel colors seemed overwhelming and extraordinary, and I grabbed one pastel after another, impulsively reacting to the scene in front of me and the box of colors beside me without such verbal questions as, "Is that blue too bright?" Choices of color were based on recognition, not verbal thought. Similarly, when an artisan is making something in wood or metal, although some actions, such as use of a hammer, are automatic and do not involve deliberate thought, and although other actions, such as checking measurements, certainly require use of words, many other tasks involve nonverbal responses. Many, like the choice of a colored pastel, depend on recognition.

The position of a nineteenth-century engineer designing a machine or structure was distinguished by the need to use drawings for communication. After James Nasmyth had the idea for his steam hammer and saw it clearly in his mind's eye, he "rapidly sketched out" the details in his "Scheme Book" so as to "render them visible." There was no process of translating from visual to verbal ideas. It was merely a matter of making visible what was already in his mind so that there was something to show to the other people who would have to help build the machine.

By contrast, craft workers or artisans making farm wagons or boats of traditional design used few drawings—sometimes none—and did not have much need for verbal explanation either. Images in the mind could seemingly be translated into physical objects. Verbal thinking was bypassed, as in my experience with the oil pastels. In a similar way, there are the mathematicians previously quoted, who show a surprising awareness of nonverbal or visual processes in their own thought, but admit to difficulty in translating them into words or equations.

Another way of understanding what is involved in nonverbal thinking could be to note the suggestion that in childhood, thought is more strongly visual than in later life. A medical man writing about art and science notes that babies develop visual skills long before they learn to talk, and a psychologist observes that children have greater "imaging ability" than adults.[26] It is reported that, as a child, Einstein was slow in learning to talk; that he delighted in diagrams and models; and that he seems to have experienced a general "displacement . . . of skill to the visual area."[27] Individuals known to me whose visual abilities seem very marked include one who mentions having been discouraged from talking much in childhood, when she was expected to be "seen but not heard." The erosion of images by words that occurs in most children as they grow up was apparently lessened for her as a result, and early visual skills persisted to a time when they could be developed in a branch of science that depends heavily on observation.

Similarly, an activity commonly noted in the childhood of engineers is an interest in making toys, sometimes accompanied by difficulties with the more academic kinds of education, indicating that these also are people with a displacement of skills to the visual area. James Nasmyth wrote that he learned little at school apart from friendship, but devoted his leisure time to making tissue paper balloons, kites, and spinning tops.[28] From such experience, it would seem that visual thinking as a nonverbal process may belong mainly to childhood, and as we get older most of us are educated out of it. A few people who retain some of this skill into adult life may then seem especially gifted in mathematics, engineering, certain of the sciences, and (presumably) art.

Perhaps because of the traditional bias toward the literary and verbal in our culture, many people are ready to assume that animals cannot think because they do not use words. Yet anyone who has watched how a cat pauses and how it uses its eyes before making a difficult jump, or who has considered what a chimpanzee does in working out how to use a stick as a banana-collecting tool,[29] realizes that some complex mental process is going on. It may not be thinking as we usually know it, but it involves a lot of visual estimation, and is too deliberate to be regarded as a reflex response, or an "automatic" action based on habit or instinct. So if, in human terms, visual thinking seems to be a heritage from childhood, in evolutionary terms, it may be part of our animal background.

Visual Skill and Human Purpose

One further point, which at first hardly seems relevant, was brought out by Anne Roe in her study of eminent scientists. She noted that several were loners, and many had acquired the habit of concentrating intensely on their work or interests from an early age. They were often ill at ease in social activities, even to the point of seeming uninterested in other people.[30] These traits showed up not only in interviews but also in test situations, in which the scientists avoided reference to anything emotional, or indeed, anything connected with interpersonal relations. All the scientists in Roe's sample were men, and she implied that unemotional attitudes expressed the "masculine image" of their subjects.[31] Other psychologists have said that although such people usually have an adequate personal adjustment for a socially useful life, some seem to form closer relationships with the objects they study or use than with other people.[32]

Roe's detailed findings are often quoted in a simplified form according to which the outlook of scientists, or nowadays more often, technologists, is described as "object-centered" rather than "people-centered." This is really a caricature, reducing the issue to a single comparison or dualism, but it is useful in pointing up certain attitudes commonly found among engineers, scientists, and today, computer enthusiasts.[33] In that spirit, object-centered attitudes or their converse are referred to again in later chapters.

For the moment, though, we are more concerned with another contrast toward which previous paragraphs have been leading, that between nonverbal, visual thinking and its verbal counterpart. Some of Roe's comments suggest that verbal skills are often more fluent among individuals with people-centered interests, and nonverbal habits of thought (especially visual thinking) may be characteristic of individuals whose outlook is object-centered.

This makes sense because words are the chief means of communication between people. And given that the engineer or scientist needs to have a practical concern with objects that often depends on visual study and measurement, verbal skill may often be less important for him or her than for individuals whose work involves frequent communication.

At the same time, though, certain visual skills are also important for interpersonal communication, notably recognition of faces, awareness of body language, and understanding of people's facial expressions (and what is conveyed by their eyes). Therefore, the rather crude contrast between object-centered and people-centered temperaments may reflect different kinds of visual skill or different ways of interpreting visual information, rather than any sharp contrast between visual and verbal ability.

Part of the difference may be a question of whether one is willing to acknowledge the aesthetic or emotional significance of visual information. Craft workers may have object-centered interests in that they have close relationships with the objects they handle and examine in their work, yet this may be an intense, participatory or involved kind of outlook, marked by the immediacy of sense experience, and by visual excitement and aesthetic pleasure. By contrast, when a scientist seems more interested in objects than people because of an emotion-avoiding strategy, he may be reluctant to acknowledge aesthetic pleasure. He (rarely she) may feel distrustful when such emotion comes to him, and may regard it as better to disregard such feelings and cultivate the detached attitude seen as proper to science rather than the participatory approach of the craft worker. So although discussion in this book often refers to contrasts between object-centered and people-centered outlooks, there are times when this dualism is too crude, and it is necessary then to distinguish participatory from detached styles of object-centered work. It can also be claimed that the training and social norms of many modern professions, even among nonscientists who work directly with people, show an increasing tendency to encourage object-centered attitudes.[34]

Recent research on autism, however, has suggested an entirely different explanation of why some individuals develop nonverbal abilities and object-centered interests. Children with this condition, who are more often boys than girls, show little interest in ordinary human relationships, and may even dislike overt affection. They are often slow in learning to talk, or do not learn at all, yet sometimes become gifted artists or musicians. Autism is known to run in families, and it now seems that autistic individuals are found more often than expected in families that have also produced engineers. Researchers comment that a parent who is not autistic may well have some of the same "cognitive characteristics or 'thinking style' as his or her autistic child," including "spatial visualization skills," and "affinity with physical objects." Often, autistic children are also "strongly numerate, recognizing pattern and order in numbers."[35]

The research on autism had not reached any conclusion at the time of writing, and it is uncertain whether the suggested link with engineering skills will be confirmed. All the same, research on the subject has thrown up many interesting sidelights on human visual ability. One point is that humans (and some higher animals) have specific capabilities for detecting purposive movement by other creatures, of monitoring the "eye direction" of other people (or animals), and of interpreting what another person is looking at.[36] High levels of sensitivity in these skills enable us to interpret body language and the "language of the eyes" and to deduce a good deal about another person's intentions. And these are some of the abilities that may be impaired in a person with autism. Such a person may have excellent visual and spatial abilities when dealing with inanimate objects (or drawings), but may miss most of the visual clues necessary in social and personal relationships, and may not be able to understand other people's intentions or purposes.

This research not only adds considerably to our knowledge of the range of visual skills but also has implications for the way we understand science. In the previous chapter, we noted that one of the advances in understanding associated with the scientific revolution of the seventeenth century was acceptance of the idea that nature can be better understood by regarding all its processes as mechanical rather than animate, and then by explaining everything in terms of cause and effect. It is worth comparing this attitude with the way some people with autism have to look at the world, as a result of impaired ability to recognize purpose in human (and animal) behavior. The disability is not, in itself, a problem when it comes to studying science. Some people with autism are very good scientists—one much-quoted case study concerns somebody engaged in agricultural research at a university in Colorado—and the biographies of some scientists (including Einstein) suggest a degree of autism in their early development.[37] One might even say that to practice science in the spirit of the scientific revolution, it is necessary to adopt a self-imposed autism for as long as one is working in the laboratory or at the computer.

The Longer Historical Perspective

Historians of science have only recently become willing to recognize a visual component in scientific thought, but are still cautious in what they claim for it. They appreciate that Renaissance inventions relating to pictorial perspective and drawing to scale assisted in the development of a new conceptualization of space. They note that evolution of visual languages for geology, chemistry, and biology in the eighteenth and nineteenth centuries allowed new concepts to evolve.[38] But they often assume that visual imagery served merely to extend the scope of verbal reasoning.

However, previous paragraphs should have put us in a position to view history rather differently. For a start, the changes that occurred, first with the Renaissance and then with the scientific revolution, can be seen as a shift from a focus on literary learning to a more strongly visual understanding of the world. This shift in focus represented a change in mind-set or *mentalité*, and we may surmise that it gave greater scope to some kinds of human ability that had been underused in the past, including the abilities of people whose emotional makeup or autistic predisposition fostered particular skill in dealing with the visual world.

This changed mind-set also made it easier for people interested in science to pursue parallel interests in engineering or other technologies and for the visual skills of artisans to make a contribution to more academic kinds of learning. My own awareness of what this might mean arises from thinking about architecture as an expression of undercurrents in both science and engineering. Thus during periods of energetic evolution in architectural ideas, such as the Renaissance in Italy and the sixteenth and seventeenth centuries in England, we can find visual skills evolving rapidly in many fields. An outstanding example is Filippo Brunelleschi in Italy at the end of the fifteenth century. His work embraced not only architecture (the cathedral dome in Florence) but also engineering (cranes, clockwork, and structural design) and innovation in perspective drawing. With regard to the last subject, there is a crossover between artisan interest and the concerns of men of learning, because although geometrically precise perspective was pioneered by artisans or people close to them (Piero della Francesca as well as Brunelleschi), the first book on perspective was written about 1450 by a scholarly upper-class man, Leon Battista Alberti.[39]

Another interest shared by Brunelleschi and Alberti was study of the surviving monuments of ancient Rome, including the great temples remaining from its pre-Christian past. To understand these buildings, both men thought it necessary to measure them precisely because they felt that the key to design was proportion. This led to a need for scale drawings, something missing from earlier architectural practice.[40]

A little later, Leonardo da Vinci was part of the same movement in the development of drawing, and his famous notebooks, full of sketches of mechanical ideas, show him to have been a visual thinker of the same caliber as the nineteenth-century engineers quoted earlier, Joshua Field and James Nasmyth.

Among those who participated in the visual/mathematical culture of northern Italy half a century after Alberti had written his important books on perspective and on architecture was Nicholas Copernicus. As a young man, he had first studied at his local university in Poland, and then came to Italy to pursue further studies in medicine—and astronomy also. When eventually he came to put forward his new view of the universe, he used an architectural metaphor that could have been borrowed from Alberti's ideas about designing churches. He wrote of the planetary system as a beautiful "temple" (the word Alberti would have used), with the sun enthroned at the center, and so placed that it could illuminate the whole at once. That this was a matter of visual harmony is clear when we find Copernicus repeatedly saying that "in this arrangement, the marvellous symmetry of the Universe" was clear.[41] Among historians who recognize some inspiration for this in Italian ideas about architecture, the only disagreement is whether Copernicus was really thinking of Alberti's writings on the subject, or of some newly built church he had seen in Italy, perhaps at Prato.[42]

Alfred Crosby, historian of ideas rather than just of science, has gone much further than most in recognizing the significance of these developments. He claims that there was a "shift to the visual" at this time in many aspects of European culture, and that it had the effect of "striking the match" that set the scientific revolution ablaze.[43] In other words, he claims that "visualization" was a major component in the new mode of scientific thought. And again using architecture to provide a point of reference, we may note that many leading figures of the scientific revolution in England were active at some time in the design of buildings. Individuals who exemplify this include Bacon, Savile, Hooke, Power, and of course, Wren. Their visual skills were applied to microscopy (Hooke, Power, Wren) as well as in more obvious areas such as astronomy (Savile, Wren).[44]

A more general point about the significance of visualization for the scientific revolution is that the distinction between objective and subjective experience became a distinction between what could be seen and what could not. Alfred Crosby links this to his point about the emergence of modern science by saying that visualization, together with objective, quantified measurement, was used to "snap the padlock on nature" so that reality was "fettered." Even music, he says, could be interpreted visually after musical notation had evolved. Polyphonic music was written that could be fully appreciated only by reading it from the page: no ear could "comprehend such complexity in time."[45] Thus when major figures in the scientific revolution such as Galileo (and also Kepler, Descartes, and Huygens) turn out to have been interested in music, we should note that their bias was often to "snap the padlock" on music as well, if not with visual representation, then with their mechanical studies of vibration. Even Kepler, who sometimes took a musical view of nature so literally that he imagined the planets singing in their orbits, spoke of the planetary system as a "heavenly machine . . . a kind of clockwork."[46]

Music and a New Synthesis

Twentieth-century science and technology have been moving away from the habits of visualization implied by the drawings done by nineteenth-century engineers, geologists, and chemists. More abstract, analytical methods are now prominent in education and practice. With regard to engineering, Ferguson sees many disadvantages in today's less visual approach.[47] In physics, though, less rigid kinds of theory have emerged. Leonard Shlain argues that artists have regularly challenged conventional assumptions about space, time, and light. They have explored spatial relationships in advance of more deliberate scientific study. And Shlain points to innovation in nineteenth-century painting that may have pre-pared people's minds for the new insights offered by relativity and quantum mechanics, however abstract they may be.[48]

This argument is not entirely convincing as history of science, but it may help explain the lives of individual scientists or technologists who play music, compose, or less often now, paint pictures. These activities may subconsciously enable patterns occurring in an individual's work to be explored more freely than is possible in a scientific context. That preliminary exploration may then allow the practical, technical work to proceed more smoothly. In many respects, as science has become more analytical (i.e., mathematical), the characteristic artistic interest of scien-tists, like that of engineers, seems to have moved from pictures to patterns and rhythms, and from the visual arts to music. In her studies relating to programmers and other computer specialists, Sherry Turkle notes that although many are musical, they are more likely to prefer the structural intricacies of counterpoint, and tend to dislike more emotional music.[49]

Generalizations about a shift of interest among technologists from the visual to the musical need to be heavily qualified, not least because of the many new kinds of visualization modern technology has made possible. Scanning electron microscopes give us new ways of picturing the structures of crystals, cells, and even molecules. Computerized imagery from space probes gives us a new understanding of many issues in cosmology. The internal organs of the human body and the mineral resources of the earth can all be scanned, mapped, and in other respects visualized in radically improved ways.

There is not only this ability to see better but also the possibility of conceptualizing many things differently. The graphs, sections, blueprints, and models of classical science and engineering all provided static images for thinking, whereas the new technologies give us moving images, helping us toward more dynamic theoretical constructs. Thus whereas James Clerk Maxwell in the 1870s and weather forecasters more recently could envisage some of the principles of what is now known as chaos theory, the visual representation of fractal patterns on computer screens has made many new insights possible. A German research team has commented that pictures have played no small role in the development of ideas, and that computer graphics, as a tool of "experimental mathematics," makes complex relationships accessible to the "intuition."[50] Here, intuition refers to thinking of the kind described earlier in which ideas emerge from visual experience without intermediate stages of verbal reasoning—ideas that can then be discussed verbally or tested using mathematics or experiment.

In this new branch of science, we can begin to see the limitations of earlier scientific thinking, with its hard-edged mechanical models. As ever-extending and often beautiful patterns are seen on screen and grow from minor and random perturbations, they challenge old assumptions about relationships between cause and effect. The German research workers have commented that even if we accept the validity of simple causal laws, and still believe that future events are determined by identifiable causes, "the predictability of the future does not follow." The tiniest disturbances at the beginning of a process can "cause completely different behaviour after long periods of time." Causality is still assumed, then, but it is a version of causality that gives rise to apparently spontaneous unpredictability. Moreover, this view offers also insight into the irregular forms we see in nature—in mountain scenery, coastlines, and clouds. Yet it is less revolutionary and more provisional as a scientific theory than some of its proponents claim.

In chapter 1, we discussed the view that any ultimate theory of nature, appropriate for a science fully at one with reality, would be so abstract and subtle that it would be nearly impossible to use. Effective science has to clothe theory with words and mathematics, with constructs such as "causality," and now we can say, with visual imagery. The descriptions given by some scientists suggest strongly how tenuous their first ideas about a new hypothesis may be. Anne Roe interpreted some of what scientists told her about their thought processes by calling it "imageless thinking." The scientists seemed to be describing thoughts that were so lacking in form that neither words, nor visual images, nor "kinesthetic" impressions could capture them. Einstein, we saw, said that some thoughts came to him as vague "muscular" feelings as well as "more or less clear visual images."[51] The struggle of such scientists to put their insights into words, numbers, or equations is evidence of the tenuous and abstract form of insights into the ultimate.

Anne Roe commented that some modes of thought she found among the scientists she interviewed were alien to her, and some seemed characteristically masculine. Other women psychologists have also been quoted here on the same subject: Margaret Shotton and Sherry Turkle.[52] And it seems clear from their work, and other evidence quoted, that what is peculiar about "masculine" science is not visual thinking as such— many women are very skilled at this—but rather the emotion-avoiding, object-centered thinking strategy and the quasi-autistic outlook of some—perhaps only a few—male scientists. Other oddities of (mostly male) experts are nicely caricatured in the fictional writings of Douglas Adams, but where ultimate knowledge of nature is concerned, Adams is serious in his belief that number lies at the heart of everything. Yet in illustrating this point in the context of chaos theory, he does not emphasize the mathematics of number, nor the visual patterns of fractals, but instead suggests that "the closest . . . human beings come to expressing our understanding of these natural complexities is in music."[53]

However, I prefer to say that, if mathematics is the best available means of handling insights into the tenuous structure of reality, music brings out some complementary aspects of reality that mathematics would sometimes miss, spontaneity and purposiveness included. Music also speaks more directly about how the abstract fundaments of nature emerge in human experience. Visual experience can capture some of these same impressions, and for me is often more vivid. But visual art and the drawings and diagrams of scientists were, for a long time in history, too static for more than limited use. Now, it seems, the computer has extended the range of visual expression into areas with which only music could previously cope.

3

Meaning in the Hands

Encouraging Invention

Creative people often speak of ideas coming to them suddenly and unbidden, usually when they have been working on a problem for some time without much success. Perhaps they have turned aside to do some other job, or else they are relaxing, and all at once the elements of a solution come to them. Sometimes this is called the "eureka effect," because *eureka* is what Archimedes is supposed to have shouted triumphantly when an idea came to him as he got into his bath. The story is said to illustrate only the point about ideas coming unexpectedly, but it actually demonstrates another effect as well. The idea that came to Archimedes was the result of an observation he made as he got into the water and noticed the level rise as his body displaced its own volume. Then he saw that the problem he was working on—how to measure the amount of gold in a king's crown—could be solved simply by immersing the crown and measuring the volume of water displaced. So there are two questions to deal with here: ideas that come suddenly, and the significance of observation.

One teacher of engineering design who appreciates the importance of sudden inspirations is Gordon Glegg. He advises that "the secret of inventiveness is to fill the mind and the imagination with the context of the problem and then relax and think of something else for a change." Relaxation, Glegg goes on, releases mental energy, "which your subconscious can use to work on the problem." Sometimes, the subconscious will "hand up . . . a picture of what the solution might be." If that happens, it often occurs suddenly and unexpectedly. This is experienced in all kinds of creative work, whether in engineering or not.[1]

It is significant that Glegg writes of a picture being handed up into the imagination, because many people who have described these experiences imply that the ideas they get are predominantly visual. Glegg goes on to quote fifteen historical examples of inspiration coming suddenly and unexpectedly. In six cases, the person concerned was half asleep or sitting quietly by the fire. In another six cases, the innovator was walking, riding, or traveling. Others were listening, perhaps not attentively, to speeches or sermons. A further example to add to Glegg's list is of August Kekulé, the chemist, who was half asleep by the fire when he arrived at the concept of the benzene ring, which revolutionized nineteenth-century thinking about molecules. The idea came to him not as a dry chemical formula but as a picture of molecules "twining and twisting" like snakes, and "one of the snakes had seized its own tail." Another idea had come to Kekulé earlier when he dozed off while riding on a London omnibus.[2]

Mathematician Henri Poincaré also admitted to experience of this kind. In one instance, he had spent fifteen days attempting a particular mathematical proof. He had tried many different angles and gotten nowhere. Then one night, after drinking black coffee, he could not sleep: "Ideas rose in crowds: I felt them collide until pairs interlocked." By next morning, the problem had been dealt with and "I had only to write out the results." In going on to the next stage in this particular research, his thinking again got stuck, until he took time off for a geological excursion. Then, when he was getting on a bus, as "I put my foot on the step the idea came to me."[3]

Creative ideas come to engineers, scientists, and artisans not only when they are waking from dreams or distracted by excursions, but at certain times when they are actively participating in work. For a carpenter, blacksmith, seamstress, or cook this may happen most particularly in the process of handling the materials with which they work—that is, when mental cogitations are interrupted by physical action. Alternatively, ideas may come when one is simply looking around the site of proposed works or checking the drawings. The attention one then gives to observation seems to free the mind for new ideas.

Apart from that consideration, observation is a visual skill that engineers need as much as craft workers. Ferguson points to cases in which reliance on computer-run checks for a design has led to important points being missed that an observant engineer would have noted. He indicates that when engineers learned drawing, this was one important way of fostering observational skill, just as botanists (for example) recognize that making the effort to draw is often the best way of learning to observe the distinctive features of plants. Engineers in training could also be taught observation more directly, and up to the 1960s were expected to examine things that other engineers had designed: "to look at them, listen to them, walk around them, and thus to develop an intuitive 'feel' for the way the . . . world works."[4]

The Sense of Form

Observation merits better recognition as a high-order visual skill. What it involves can be illustrated by considering how people may look at a landscape and see only its gross features: fields, roads, forests, mountains. There may be irregularities in the layout of some fields, and bumps or depressions in the land surface, but these are disregarded as minor and apparently random occurrences. To an experienced observer, however, such irregularities may show a significant pattern underlying the superficial grid of modern fields and roadways, a pattern that may speak volumes about rock formations, or about how land has been eroded by water or ice, or about ancient man-made features. Charles Darwin once described how he missed features of these kinds in a valley in Wales that he visited with geologist Adam Sedgwick. They were surrounded by evidence of how the valley had been formed by the action of a glacier, including perched boulders and terminal moraines, "but neither of us saw a trace of the wonderful glacial phenomena." At the time (1831) ideas about ice ages and glacial activity were still unfamiliar and even an experienced geologist had no sense of what to look for, or how to recognize what he saw.[5]

The ability to recognize patterns of one kind or another is important in a variety of disciplines, and may be compared with the ability of a good engineer to evaluate a structural design "by eye." What is needed in every case is a sense of form derived from an accumulation of visual memories of how particular structures may look in a great variety of contexts, both adverse and favorable.

Michael Polanyi provides an example of how this works in discussing the experience of a medical student learning about the X-ray diagnosis of pulmonary disease. "At first the student is completely puzzled." All he or she can see in the X-ray photographs of a chest is "the shadows of the heart and the ribs, with a few spidery blotches between them." Radiographers who point out other details seem to be making it up. But after seeing pictures from many different patients, the student will "forget about the ribs and begin to see the lungs," rather as a student of landscape can forget modern fields and begin to see the geology or archaeology underlying them. Eventually, a rich panorama of significant detail is observed, including (on the X-ray picture) detail of "physiological variations and pathological changes, of scars, of chronic infections and signs of acute disease." Some people would say at this point that they have "got their eye in." Polanyi's comment is that the student "has entered a new world."[6]

That happens only after the student has looked at many pictures, accumulating memories of patterns that comprise a sense of form. This does not mean that every X-ray image is remembered individually and consciously compared. The process operates to a large extent intuitively and unconsciously, partly because one's visual memory stores more information than can be recalled item by item. Similarly, a working engineer, builder, or carpenter picks up many visual impressions that do not get consciously labeled, yet are remembered in a general way and are brought to bear on subsequent jobs. Designing a bridge, an engineer may deliberately refer to features of other bridges, but less-specific memories also come back as work proceeds. Some of these memories may not emerge into full consciousness but have their influence through hunches and intuitions and a sense of fitness for purpose. Polanyi refers to the accumulation of memories, hunches, and also unconscious skills as "implicit" or "tacit" knowledge.[7] Much of it is visual knowledge constituting the sense of form. In technology, that means form and pattern appropri-

ate for various functions, given the materials used. Traditionally, craft
workers depended heavily on this kind of knowledge as distinct from the
explicit, rational methods of modern disciplines. However, engineers are
still likely to make use of tacit knowledge to a considerable extent, and
probably more than they would want to admit.

Douglas Hofstadter, a specialist on artificial intelligence, offers a useful
comment on this phenomenon when he talks about a "feeling for form"
in mathematics. The problem with deductive reasoning, Hofstadter says,
is that it leads to too many conclusions. Choices have to be made as to
which conclusions are significant, and then judgment is called for. Tied
up with this judgment is intuition, or a "sense of simplicity," which
Hofstadter equates with a "sense of beauty," and "an elusive sense for
patterns."[8]

I would differ from Hofstadter only in emphasizing more strongly that
the sense of pattern to which he refers is not so elusive that it cannot be
educated, as in the training of medical students with regard to X-ray
pictures. In craft skills and engineering, what is also important is the
accumulation of memories of what has worked well in the past, so that
a sense of form develops that is especially discriminating about soundness
of design of bridges, boats, aircraft, or some other speciality. Aircraft
designer A. V. Roe (of the former AVRO company) is said to have taken
his holidays on the coast each year, choosing places where he could sit
on cliff tops and watch seabirds wheeling and gliding.[9] Visual memory
tends to be vague unless one has been trained in observation. Spending
a long time just looking like this could add much detail to Roe's sense
of forms suited to flying.

Henri Poincaré spoke of "mathematical beauty" as a "true aesthetic
feeling that all real mathematicians know." References to aesthetics in
mathematics, aircraft design, or engineering are often indirectly referring
to the individual mathematician's (or engineer's) sense of form, and
denote the pleasure and completeness experienced when this sense
matches a proposed solution to a problem.

It has often been said of Robert Maillart, pioneer of reinforced concrete
bridges, that his new structural forms arose out of aesthetic feelings as
well as from scientific ideas. It appears that Maillart relied heavily on an

aesthetic feeling for form in developing his designs, and that he carried out structural calculations only afterward, to check that his intuitive ideas about the best shape for a particular bridge were correct. Maillart admitted that on occasion, he was slightly misled by memories of masonry bridges.[10] When he began his career, there were so few concrete structures to look at that his sense of form was not at first sufficiently educated to recognize how light and elegant a reinforced concrete bridge could be.

Gordon Glegg asserts in general terms that "an engineer . . . is a creative artist in a sense never known by a pure scientist. An engineer can make something." Robert Maillart's bridges are widely recognized as artistic achievement of a high order, not just triumphs of engineering. Indeed, the most prominent commentator on his work suggests that we should recognize that "structural art" is indeed an art form, just as surely as architecture or sculpture.[11]

Although I share that sentiment, my purpose here is to stress that a great deal of work in technology feels like art to those involved, even if it does not communicate expressively with other people as conventional artistic work is expected to do. It feels like art because it so regularly calls for aesthetic, quasi-intuitive judgments. So often a car, boat, or bridge that has been well designed for its function is aesthetically satisfying to look at, not because functional shapes have an inevitable beauty, but rather because aesthetic judgment was used in achieving an effective functional form.

A final point about sense of form is that it plays a part in many other activities, in which, as in technology, it is developed by habits of observation. An example is the work of Barbara McClintock, winner of a Nobel Prize for work on the genetics of maize. Her biographer comments that seeing in science is like seeing in art. It is not just objective, but depends on relating what is seen to a pattern, form, or vision in the subconscious. Using a microscope to study cells from maize plants, McClintock felt that she was able to get right down inside the cell and look around, and her discoveries came not only from direct observation, not only from this particular use of imagination, but also from her "internal vision," which is how McClintock described her sense of form as it applied to cells and the chromosomes within them. Her vision, she said, gave her joy, and although she was not designing artifacts (as Glegg would point out), she did have to design experiments. And about them she remarked: "When you have that joy, you do the right experiments."[12]

In bird watching, similarly, one acquires visual knowledge of a particular species from repeated sightings, until after a time it is possible to develop an eye for that kind of bird. Then it can be recognized at a glance, or when seen only partly or briefly, or when it is moving rapidly, by its style of flight. Some bird-watchers talk about the "gizz" of a bird, meaning various telltale features of its appearance and behavior that distinguish it from other species. One's visual memory stores far more information than can easily be recalled and handled in words, as in the experience of engineers, builders, and carpenters.

Using All the Senses

Frequently today there are complaints about neglect of training in traditional visual skills, especially where they once involved drawing. That seems especially ironic in view of the many advances in technology that are increasing the quantity and quality of visual information available. For example, with the aid of digital image processing of signals from space probes and telescopes, astronomers are able to see more distant objects and in greater detail. There is a real exhilaration about the work in which sheer visual pleasure plays a part.[13]

One problem seems to be that the use of such equipment is sometimes associated with a narrowing of the focus of what one looks at. The equipment itself is part of the fascination, so one tends to look only at what it reveals. An account of advanced sidescanning radar used to map the seabed notes that the scientists responsible were so involved with the technology that they failed to observe many aspects of the marine environment that they could have seen from the deck of their ship and whose relevance they should have appreciated.[14] Another dimension of the problem was one I encountered when being taught a course on soil science in the classroom and through textbooks. I became increasingly baffled until, one day, we were taken out into a field and were asked to sample the soil using augurs. A pit was dug so that we could see the soil profile, and we were shown how to feel the soil between our fingers, thereby judging its sand, silt, and clay content. This was real observation, not pictures on a screen, nor textbook diagrams, but something literally tangible, and the subject immediately became alive and acquired meaning for me.

Tracy Kidder has remarked that for some engineers, "reading does not constitute knowing. For them touch is the first of the senses." So to understand a computer, some engineers might take it to pieces and handle its printed circuit boards. Of course, "form on the surface of a board" does not say much about how the computer works. What it does is to make abstract knowledge of that kind seem real and within reach.[15]

These examples show that many technical subjects gain an extra dimension for some people through seeing and touching as well as knowing. But having said so much in the previous chapter about visual thinking and in this chapter about visual observation and the sense of form, I now need to stress that all the senses, not just seeing, can enter into thought this way. Chemistry depended on such craft experience from the time when it was undistinguishable from the work of apothecaries, assayers, and alchemists right through into the middle twentieth century. When I was introduced to laboratory chemistry, observations of color changes and smells were strongly emphasized, and textbook writers still consider it important to say when a substance has a characteristic smell. Elderly chemists have said how they used taste as well as smell in somewhat hazardous analytical tests. Metalworkers, according to Cyril Stanley Smith, regularly developed new skills through aesthetic response to sense experience. He was thinking mainly of visual observation of the sheen, texture, and "watering" of metal surfaces, but accounts of early technology also mention sounds, such as the ring of metal on a smith's anvil.[16]

In many other jobs, even in writing (according to Boris Pasternak), "the living movement of his hand" and the rhythm of the work may help the author or craft worker maintain the flow of what is being done.[17] Many handicraft processes are slow, repetitive, and tedious, yet require precision and hence concentration. One thinks of sewing a garment, working a piece of stone with a chisel, or shaping metal on a blacksmith's anvil. The musicality of these tasks allows the craft worker to cope with repetitiveness and avoid loss of attention due to boredom. In one instance, "the rhythm became like a pleasure-giving drug," and that alone made it possible to keep going.[18]

In preindustrial circumstances, much practical knowledge may have been gained through the hands.[19] In paper mills, boatyards, and potteries, products were made without much reference to such purely visual forms as drawings. Workmanship depended on handling materials as well as on vision, and one philosopher of technology has noted that in some senses, artisans may have been "thinking with the hands."[20] What this could mean is indicated by the way one potter described improving designs by working manually with her materials: "eyes and hands will help you make better pottery than any theoretical analysis of form," she said, adding that she learned more by "seeing and feeling" for herself than from instruction.[21]

Similarly, in building farm wagons, a wheelwright knew about the timber he worked with, "not by theory, but more delicately, in his eyes and fingers." His arms learned "to recognize what mattered" about the weight of a plank, and when it was being sawn, his eyes would watch for "any hint of pinkness—a fairly sure sign of sap," or his "nose might detect the sappy smell." Paying attention to color and smell and what one's fingers felt was thus crucial for making technical judgments, but it was more than that, because such experience became a vivid part of what it felt like to be a woodworker or potter. It could endow a fairly ordinary job with meaning, and could give the wheelwright a lift as he worked. Looking back, George Sturt exclaimed: "Lovely was the first glimpse of the white ash-grain, the close-knit oak, the pale-brown and butter-coloured elm." Another wheelwright especially remembered a sound: "the hiss of ash being shaved" (or planed), plus a tactile experience: "the satisfaction of running one's hand over a nicely shaped spoke."[22]

In all the woodworking trades, such experiences of sound, touch, and the smell of wood were coupled with more abstract knowledge in making choices of timber for different purposes. In an English timber-framed house, oak might be employed for the exterior where elm was used inside, and there was need to select curved timbers for some purposes (e.g., wind braces). Which piece of timber to use where was often a matter for careful judgment.

Wheelwrights consistently used split oak for spokes of wheels so there would be no cross-grain, and elm for hubs. Ash was widely used for the felloes (or rims) on which the iron tires were fixed, because it was the most elastic of any wood, although in southern England, beech might be used. The floor of the wagon body was often made of elm, whereas the sides were of oak. Similarly, before iron was an option, ash was normally used for making the moldboards of plows, whereas among millwrights, apple wood was regarded as ideal for the teeth of gear wheels.

Tactile as well as visual judgments were involved in the selection of timber. The wheelwright would pick up a sawn length and spend some time "twisting it, turning it end-for-end." He would know "where a hard knot may even be helpful and a wind-shake a source of strength."[23] In other craft technologies, a larger or different range of both visual and tactile experience may be involved. For example, in working up clay to make a pot, what the worker feels with her fingers is crucial.

In metallurgy, Cyril Stanley Smith has argued that experience and skill are greatly enhanced when craft workers have opportunities to make decorative or artistic objects. Their repertoire of skills remains limited for so long as they work only on utilitarian products. When the design and decorative treatment of swords, pots, or textiles becomes an art, and not merely the production of useful articles, the craft worker is brought "into contact with more properties of materials than are encountered by the maker of useful objects."[24] However, even when the purpose is utilitarian, making something in metal, wood, or clay is always likely to be a significant visual and tactile experience. The changing colors of iron in a furnace and on the anvil are the same—as are the grain, texture, and scent of wood—whatever is being made.

Many similar points can be made about mechanical or electronic equipment. Maintaining a motorcycle, a good technician often diagnoses where there are problems without going through the full fault-finding routine. He or she is able to bypass parts of the testing procedure by picking up small clues—audible as well as visual—that give a feel for what may be wrong.[25]

A similar ability was noted in men who repaired aircraft radio sets during World War II. In the urgency of war, they were given little training but gained much experience. All they could do at first was follow a routine testing procedure until they found a faulty component. But in time, many developed an uncanny knack of finding the damaged component without making all the tests. In other circumstances, it was noted that some who had this unusual sensitivity for radios showed it in the "strong, careful handling" of equipment and in "the gentleness of the true craftsman."[26] They still did not understand intellectually how radio sets functioned, but by working with them, they had developed a feeling for the kinds of fault that most often occurred and the symptoms that resulted.

Needless to say, the sights and sounds a mechanic or radio technician encounters, or a blacksmith working iron on an anvil, can awaken strong aesthetic feelings. Whereas some reflect qualities in the material being worked, others are related to the process of making technical judgments. When there is a good match between what a blacksmith observes on the anvil and his or her sense of form, this is understood, not by doing calculations, but through feelings of recognition and aesthetic satisfaction.

In England during the seventeenth and eighteenth centuries, widows quite often carried on their deceased husbands' businesses. Women blacksmiths and bronze founders (such as Jane Brewer, whose foundry was in St. James's, London, in 1707) may have been rare, but there were many women silversmiths whose aesthetic achievements show evidence of the same enjoyment of materials as described for other metalworkers.[27]

Artists and Alchemists

Earlier in this chapter, one aspect of meaning in technology was explained by saying that, for many practitioners, engineering feels like a creative art. Onlookers may not perceive any art (although they do for the work of silversmiths), but it may still feel like that to someone making things in a process informed by aesthetic responses. It now appears that some of the meaning of technology for those most directly involved comes from experience of handling materials: soil, timber, metal, or even radio parts. And that suggests another way of explaining what is at issue, namely, by comparing the experience of craft workers and engineers with what alchemists once talked about.

Here I am not thinking of alchemy as the old, deluded effort to turn base metals to gold, but of a very practical tradition of alchemical interest in Europe around 1600. The people concerned were artisans in metal-working trades and involved in medicine (e.g., apothecaries). What is of importance about these groups is that they were practicing science with a "participatory" attitude. The language, concepts, and methods of alchemy gave expression to the two aspects of participatory technological experience that have been emphasized in this chapter: first, knowledge derived from the senses—touch, smell, and sight—and second, intuitive discovery of the kinds represented by the eureka effect or involving the sense of form. The latter were referred to in the alchemical tradition as knowledge resulting from "illumination" and "the light of nature." Thus it was possible to acknowledge "an inspired level of experience."[28]

It is important to give attention to these ideas because alchemically minded craft workers were exploring practical experience in ways we no longer do. Moreover, their work was related to a large body of technological activity associated with mining and smelting metals, assaying ores, etching with acids, and formulating medical remedies. This was also the period of the scientific revolution which, in principle, was hostile to alchemy, but which is now understood to have made positive use of it in many instances.[29]

Alchemy involved real practical work with furnaces and stills. It led to discoveries about acids and alkalies and to improvements in equipment. As also in more ordinary craft technology, there was a great deal of monotonous work, such as grinding up raw materials or ingredients for medicines. Monotonous but rhythmic work could induce mood changes, as we have seen, and whereas for an ordinary artisan these might simply serve to make a boring task tolerable, for alchemists there was sometimes an attempt to manipulate and enhance such feelings by fasting or meditating. Then talk about transmutation of base metals into gold may sometimes have referred to elevated feelings accompanied and symbolized by some quite modest chemical change.

For our purposes, however, the most important writers in the alchemical tradition were those who retained strong links with craft technology, while also maintaining connections with formal learning. The most notable of all authors in this respect was Paracelsus, sometimes known as the "Luther of medicine," because of the reformation he began in that field.[30]

Born in 1493, Paracelsus was a German-speaking Swiss who traveled widely in the mining districts of Germany and eastern Europe. He is said to have served an apprenticeship in a mine as well as receiving a partial education in medicine. He also worked for a time as an army surgeon and taught at Basle University for a year. Little of his work was published until 1570, but then his books attracted much attention, and by 1600 were widely read. In England, a minor revolution in medical therapy[31] was accomplished as chemical and mineral remedies of the kind he had advocated were included in pharmacopoeias from 1618.

Paracelsus not only wrote in a practical way about medicine but also felt "incited" to write "a special book concerning Alchemy, basing it not on men, but on Nature herself, and upon those virtues and powers which God, with His own finger, has impressed upon metals."[32] In other works, often obscure and mystical, yet also having practical emphasis, he presented alchemy as the art of making medicines. The "steps of alchemical knowledge" were identified with laboratory procedures such as distillation. Indeed, a laboratory equipped with furnaces for such tasks was regarded as the true home of the physician, who should make up his own remedies, even if he got dirty in the process "like a blacksmith."

Sometimes, then, the Paracelsian doctor seems much like a craft worker, though at other times he is a wise man or "magus" playing with esoteric ideas derived from ancient oriental traditions. The very word "alchemy" indicates a derivation from the Arabic *al-Kimya*.[33]

During the later phases of this tradition in England, a certain John Webster wrote eloquently about its relevance for reform of science teaching in the universities. Stressing the importance of immediate experience of the senses, Webster insisted that youths wishing to learn must "put their fingers to the furnaces, that the mysteries discovered by Pyrotechny, and the wonders brought to light by Chymystry, may be rendered familiar." They should learn, he said, by "manual operation and ocular experiment."[34]

But Webster, like many English Paracelsians, was strongly religious in outlook. He served as both physician and chaplain with Puritan armies during the English Civil War. So when he came to talk about inspirational

experience in practical chemistry—arising from the eureka effect, perhaps—he expressed himself most readily in religious language. Much
conventional science was like worldly wisdom, he said, in being "meer
(*sic*) foolishness," but the more practical forms of "Chymystry" gave
access to inner knowledge—technical as well as spiritual—which comes
through rays of "Caelestial light that the Spirit of God reveals."[35]

After his retirement, Webster published a book entitled *Metallographia*
that was a solid and competent review of lead and copper mining in
northern England. He had visited mines and worked with assayers, but
was emphatic in quoting Paracelsus as the source of his inspiration, and
made no attempt to disown the alchemical background of his work.[36]

Ideas drawn from this tradition of practical alchemy influenced many
of the central figures of the scientific revolution, including (most famously) Isaac Newton, whose studies also took in mystical and historical
aspects of the subject. One way of understanding why this approach
continued to be significant is to reflect that in science, as in any of the
craft technologies, personal feelings and individual responses are always
involved, not only objective knowledge. Alchemy in its seventeenth-cen-
tury forms is not to be dismissed as the pursuit of mere illusion. It
encouraged experiment and yielded useful knowledge, and notably, it also
attempted to stay in touch with participatory experience—that is, experience of what it felt like to practice chemistry or metallurgy.

From Alchemy to Information Technology

Although people employed in laboratories, workshops, and industrial
plants have continued to use immediate experience of their senses—sound
and sight, touch and smell—and although they have continued to be
prompted by an intuitive sense of form, there has been no consistent
philosophy for this aspect of technology since the decline of alchemy.
During the seventeenth century, men of science who preferred a mechanical worldview to the alchemical one dismissed experience of touch, color,
and smell as referring only to "secondary qualities." Toward the end of
the next century there was a period of romantic reaction against this
outlook in which Goethe's theory of colors and some renewed interest
in the alchemical tradition had influence, but there was never a viable

alternative to mechanistic, masculine science. In today's world, enthusiasts for virtual reality do pay attention to sense experience, but only in order to replace it by simulations. These, according to Mark Slouka, are then often regarded as more real, and of greater value, than experience of real life.[37]

Referring to an earlier phase in technological development, Jacques Ellul remarked that man had "lost contact with the primary element of life and environment, the basic material out of which he makes what he makes. He no longer knows wood or iron. . . ."[38] With regard to iron-working and other metallurgical skills, there is a long history of craft knowledge and expertise being displaced, partly by mechanization and automation, and partly also by more scientific forms of expertise. Chemists were employed in steelworks from the 1880s to make judgments about quality, and to check the stage reached by some processes, so plant operation became less dependent on artisan skill. But many tasks continued to require traditional knowledge and judgment until very recently. One Sheffield steelworker claimed to have twenty-six methods for judging when steel is ready for cooling in one particular process, but he informed the engineers installing a computerized control system of only one of them.[39] Though his claim cannot be literally true, because one cannot enumerate intuitive ways of making a judgment like that, the story underlines the continuing relevance of some craft knowledge and points to difficulties when computer control systems are introduced. In this instance, managers expected modernization to yield improvement in reliability, economic performance, and quality control. They also expected to employ fewer skilled people. But the problem now was that plant operators sat in a control room remote from the material being processed. They were surrounded by data in the form of digital printouts and displays on glowing screens. There was just as much skill and insight needed as in using twenty-six visual-cum-tactile methods for judging the state of the steel. But it was a different kind of skill, and until it had been learned, plant breakdowns were more frequent and costs were adverse.

Shoshana Zuboff has described a similar experience at a pulp mill and paper-making plant in the United States in a discussion of information technology. As originally built in the 1940s, the pulp mill depended on operators' having an amazing range of craft skills. One man was in the habit of "sniffing and squeezing" the pulp to judge its chlorine content, and another used the static electricity in his hair as a check on the amount of moisture in the environment.[40]

Zuboff comments that the operators were controlling a complex process very skillfully using tacit knowledge (in Polanyi's sense) that "they were unable to describe verbally."[41] They were practicing a craft skill—even a form of alchemy—that depended on the immediacy of sense experience.

When a modern computerized information and control system was introduced in the 1980s, with sensors to record all the information that before had been obtained by direct observation, the same problems were encountered as at the steelworks. Breakdowns were more common, and in addition, managers realized that their new system was generating masses of information that would not be used unless more people with new skills were employed. Slowly the managers learned that the plant did not, after all, require fewer skilled operators. It had not been "automated" in that sense. Rather, as Zuboff saw it, the plant had been "informated."

In the modern world, control of industrial processes depends on interpretation of digital and similar information rather than on sense experience. But operators still need insight and imagination, and it would be surprising if they did not still depend on a sense of form and occasional eureka effects. Except in processes where the new imaging technologies are used and visual skills are needed, though, immediate visual and tactile experience is hardly relevant in the old way. An enhanced level of skill is required, often by more people, but it is skill in conceptual (verbal) thinking and the ability to perceive patterns and relationships in digital material. Zuboff sees this as giving technology greater human meaning because, unlike automation, which demands less of fewer people, these applications of information technology demand more, and require greater knowledge and insight.[42]

It is ironic that, first during the scientific revolution, and now in the information revolution, people thought they had made knowledge entirely objective and believed that they had disposed of the need for intuitive judgement based on participatory experience—yet in both instances, some of these skills still proved to be necessary. It is as if some vestige of alchemy remains in the way we practice technology. Scientists, we noted at the beginning of the chapter, may still gain ideas from dreams. A Nobel prize winner may still talk of "internal vision" playing a part in her work, even as alchemists spoke of the "light of nature."

The word "participatory," as used by Morris Berman and others, is key to the point at issue.[43] It denotes knowledge gained as a result of being personally involved in one's work (rather than being always detached), because personal involvement makes one responsive both to sense experience and to hunches (or intuitions) such as those arising from the sense of form or from eureka effects. The main difference between traditional craft work and the examples quoted toward the end of the chapter is that today, less awareness of sense experience is involved. Instead, much information is presented on computer screens and printouts, and does not require direct contact with the materials being worked. Visual skills are still required, although in a physical sense, the work is less participatory. And as the next chapter suggests, visual awareness is still an important part of human experience of technology.

4

Social Meanings

Interpreting Social Purpose

For too long, according to David Nye,[1] "it has been assumed that the social meaning of a new machine was defined by the inventor." To counteract that bias, we need to consider the meanings of technology that may be discovered by users of machinery, factory workers, or consumers, or else are imposed by corporate marketing strategies. Inventors, we may then find, do not so much define social meanings as respond to them. Indeed, their inventions and designs may often be prompted by the meanings that society has invested in already existing technologies.

This theme has not so far been discussed here—and is not destined for extensive discussion—because this book aims to explore the more immediate kinds of personal experience of technology, such as the visual and tactile experiences of materials described in the two previous chapters. On that level, nearly everybody has some experience that can be compared with the experiences of engineers, artisans, scientists, or inventors. For example, the domestic work carried out by any householder is likely to have included cooking, repairing or making clothes, putting up shelves, or mending furniture. As experience of any of these activities accumulates, one learns to make judgments based on color, texture, shape, or structural form, and one often finds enjoyment in the physical experience of working with the materials. Engineers and artisans may work more often with metal than with clothing fabrics or foodstuffs, but the experiences of learning and of enjoying materials are similar.

Some everyday experience of tools and machines is also comparable to more specialized experiences in engineering or industry. Driving a car is a craft skill that depends on sensitivity to visual information, sounds, and rhythms, and in that respect alone can be very enjoyable. However, beyond the immediate experiences of driving and domestic work, which certainly have meaning in themselves, there are social meanings related to the wider purpose of the job at hand. In cooking a meal, one is not motivated just by the colors and textures of foodstuffs, or the sensations of heat and smell that come from the oven. Nor is the technical interest of a new recipe a sufficient object in itself. Many people lose interest in cooking when they have nobody to cook for, and it is clear that the social purpose of a meal cooked for a family or to mark a special occasion is of great significance. Similarly, in discussing the motivations of inventors and engineers, we need to know about the wider social meanings of technology, and how the inventors and others respond to them.

So the discussion presented in previous chapters, however it might be extended, cannot give us the whole story of "meaning in technology." Conversely, though, discussion of social meanings, even when outstandingly perceptive, does not give the whole story either, because personal and sensual responses to technology as well as social meanings affect the way householders and car drivers use the appliances and machines they control.

Evidently, then, the *social* meanings of technology coexist and interact with the *personal* responses and "existential" experience of individuals. A cook who does not enjoy the colors, textures, and scents of food in different stages of preparation never becomes skilled at the job. But the cook is also motivated by awareness of the social purpose and context of the meals he or she prepares. Similarly, inventors, engineers, and artisans may have intense personal experience of materials, or of sweetly running machines, but at the same time, they are also members of society, responding to public enthusiasms, political influences, economic conditions, and other aspects of their social environment. The inventors of television, for example, were well aware of the social meanings of the theater, music hall, and cinema, and were aware also of the potential of combining the visual, dramatic appeal of these media with the immediacy of radio.[2] But as inventors and technicians, they simultaneously responded to all the experiences that a craft worker encounters as he or she tinkers with equipment such as cathode ray tubes, amplifiers, and scanning systems, experiences that foster a feeling for what the equipment can do.

Two levels of meaning are apparent, therefore. On one level, the inventor is informed by experience of society, sometimes mediated by economists or market researchers, and from that gains a sense of what desires or needs await a technological response. On another level, though, practical experience, visual memories, and a developing sense of form together prompt ideas about design and guide the assembly of valves, condensers, and cathode ray tubes.

Different levels of meaning are also evident when research and development are carried out within large institutions, whether industrial corporations, university laboratories, or government research stations. These can be closed societies for which a project has meanings of its own that are not shared by a wider public. Studies of life in weapons laboratories show, indeed, how research can become self-contained in precisely this way.[3] The same has been said of a laboratory where scientists inserted genetic material from a luminous jellyfish into crop plants so that the plants glow in the dark when attacked by fungus. Farmers can then inspect their fields at night and identify locations where the crop needs spraying, thus saving the expense of spraying the whole field. One response to the idea of luminous crops, as to some of the innovations of weapons laboratories, is that the scientists had lost touch with reality. "Having cut themselves off from other people's lives and from other academic disciplines, professionally blinkered scientists are unqualified to determine whether . . . they are studying something worthwhile."[4]

That comment comes from a biased source, and elicited a pained response from the scientists concerned, saying that their research would help farmers feed the growing world population while cutting back on the use of pesticides. However, researchers who make such claims are indeed often found to be cut off from the people they seek to help by their laboratory environment, and are sometimes unaware of the way agroindustrial businesses exploit science and innovation to control markets, often in remote countries. Instead, the scientists seem beguiled by the technical challenges of their work. They may wish to help farmers produce more food, but they never encounter the reality that the farmers face. If the scientists' projects are not to miscarry, it is important to ask: "Whose reality counts?"[5] Whose understanding of the meaning of technology prevails?

A less contentious example of different perceptions of meaning in technology is the nineteenth-century steam locomotive. As part of a railroad system, the locomotive could represent many social purposes connected with transportation and could symbolize the financial dealings of the company that owned it. But there was also experience of the sight and sounds of a locomotive at work, perceived differently by different individuals. The initial design of a new engine type depended extensively on visual thinking by individual engineers, though with details worked out by teams of people in drawing offices. One such office, described by Eugene Ferguson, had pictures of locomotives hanging on the walls and piles of drawings available for consultation, so that visual reference material could be frequently consulted.[6] For the engineer and his drawing office staff, the meaning of a locomotive design, logical and neat, had something of the quality of a well-proportioned classical building.[7] By contrast, those responsible for operating locomotives listened to the exhaust beat and the clicking or clanking of valve gear with a discriminating ear that could pick out inefficient steam cutoff settings or potential mechanical malfunctions, or that could foster a sense of deep satisfaction when all was running well.

Yet members of the public and train passengers would hear these noises differently. For them, the musicality of a locomotive in action would be perceived as dramatic and powerful sound rather than as ordered rhythm. Walt Whitman memorably captured this when he addressed a locomotive thus:

"Fierce-throated beauty!
Roll through my chant with all thy lawless music. . . ."[8]

An engineer listening carefully to check the performance of the machine might comment on its sweet running, but Whitman stresses the different public perception by finding "no sweetness" in the "measur'd dual throbbing" or in "trills and shrieks by rock and hills return'd." The locomotive is not here a work of classic design, but a rival to wild nature, triumphantly conquering the distances of the vast American continent.

Consumer Goods

It is clear from the last example that locomotive designers, engineers and passengers all perceived the same product in different ways, because of the different experiences involved in making or operating the machines, compared with just riding in a train. The same contrast is evident in the householder's experience of domestic technology. Making clothes or cooking a meal means being involved in a process, manipulating tools and handling basic materials, but an equally common experience in today's consumer society is to be confronted with a finished product of definite size and shape that has only to be plugged into an electricity supply and is then ready for use. In experience of a process, visual and tactile impressions are important for making judgments about how to proceed, and they contribute to the enjoyment and satisfaction one gains from doing the job. In experience of a finished consumer product, however, one is offered visual and other impressions that have little relation to practical judgments, but usually have symbolic meanings. For example, many domestic appliances have traditionally had shiny white surfaces. This has a practical purpose, of course. It helps make an appliance easy to clean, and makes the dirt that needs cleaning off easier to see. But white came to symbolize a whole range of other meanings as well, at least after 1893–94, when the Woman's Building at the Chicago World Exposition, with white Beaux Arts decor, showed an all-electric kitchen well ahead of its time.[9]

In the next few decades, a growing home economics movement increasingly focused on the need for domestic cleanliness to protect health. Devastating epidemics were remembered only too well, and infant mortality was still very high. The germ theory of disease had emphasized the connection between illness and dirt, and the message was repeated by teachers of home economics, and by advertisers of soap and easy-to-clean equipment. An example of the latter, presented as "easily kept sweet and clean," was a porcelain-lined bath manufactured in Pittsburgh, Pennsylvania, about 1900.

This all constituted pressure, not only to equip the home to promote cleanliness and health, but for women to work harder to achieve better hygiene. Because this ethos persisted through the 1920s and 1930s when domestic servants were fast disappearing, it meant that the personal experience of many a married woman was of being more tied by work in the home.[10] In *personal* terms, then, meaning in technology could have a narrowing, involuting focus, even while the *social* meaning was positive and progressive.

Although white was often used to symbolize cleanliness and purity throughout this period, it was not at first specifically associated with cookers and clothes washers. Many of these were of mottle-blue enamel or other darker colors, occasionally with white panels. Flush surfaces, wholly finished in white enamel, seem first to have been used for refrigerators during the interwar period, again to stress cleanliness and the importance of keeping food free of germs. Designed by men, and often purchased by men for women to use, they were another reminder of what men thought about women's role in the home, and in keeping the home clean.

In 1990, cookers, refrigerators, dishwashers and the like had been traded for several decades as "white goods," even though other colors were by then being introduced. It was also significant, though, that much leisure equipment had a matte black finish, including video recorders, CD players, and cameras. One study of how technology is used in the home noted that the contrast in color (reflecting domestic work and leisure) correlated with meanings regarding gender.[11] Designers had chosen black when aiming to appeal to men, and sometimes deliberately borrowed a "combat-ready" look from military equipment to symbolize no-nonsense functional rigor. At the Pompidou Center in Paris, an exhibit dealing with consumer design incorporated a jet fighter to emphasize this. However, the combat-ready men in some households seem unable to understand washing machines, and the question arises whether color and finish have anything to do with it. What would happen, one journalist asked, if white goods and black (or brown) goods were to swap colors? Would men be able and willing to use a matte black washing machine?[12] Certainly, push-button controls and digital displays seem to make microwave cookers attractive to and useable by men.

Another, much older symbolism operating on a domestic level was pointed out by Lewis Mumford when he observed that we tend to identify technology with tools, and forget that its products also include pots and pans, bins and baskets, hearths and houses. The difference between the latter, which are all containers in one sense or another, and the tools and machines that preoccupy us more often is that we use tools to do things to the world. But we use containers to conserve and protect things.[13] If people regard tools as prime examples of technology and forget containers, that says a good deal about the kinds of purposes for which they prefer to use technology. It implies that they are more interested in action rather than conservation.

Discussions that consider the symbolism of tools or containers—or the colors of consumer products—may seem unimportant except to marketing people and advertising agencies. However, those who systematically analyze the visual meanings of artifacts argue that we all need "a more adequate comprehension of one of the most important nonverbal languages—namely visual language." They claim that study of "visual semiotics" can provide an essential balance "to the one-dimensional tendency" in the conventional emphasis on verbal thought and culture.[14]

The study of semiotics, whether visual or otherwise, entails studying communication, and in a very broad way, we may think of technology as a communication system, in which inventors, designers, and builders of artifacts are "senders" of messages, and consumers, users, and the public are "receivers." Susan Wittig notes that there are several ways of analyzing a communication system of this kind[15]—one can emphasize senders, or one can focus on receivers, or one can study the messages that pass between them.

Many of the questions asked in this book are about the meaning of technology for engineers, artisans, inventors, and other senders, although questions are also raised about the messages that society sends back to inventors and designers. However, analysis and criticism can also focus on the signal and code used in communication, and we may analyze what Winner identifies as "artifact-ideas,"[16] with the understanding that every product means something. Susan Wittig, though, is most interested in a third form of study that deals with what a message signifies from the receiver's point of view. This depends on the knowledge, values, and purposes that the receiver has in mind when interpreting the message.

Variable Social Meanings

The point that what is signified by technology depends on the people
who are receiving the message is expressed in a slightly different way by
Carroll Pursell. He asserts emphatically that "the utility of a tool is never
simply in the production of *goods:* the tool also produces *meaning.*"[17]
He then adds that the meanings so produced in a particular artifact are
variable. Designers may intend one thing, but production workers, technicians, and consumers (all of whom are receivers of the technology in
one way or another) may place other meanings on a product, including
"marks of skill, alienation or masculinity."

Historians of technology of the social constructivist school have often
commented on how the same technology may have different social meanings for different groups of people.[18] Their most famous case study refers
to the bicycle, and this can be illustrated by reference to the time around
1890 when the "safety bicycle" encountered problems because its solid
rubber tires gave users an uncomfortable ride with much vibration.
Pneumatic tires were seen as a possible solution, but were regarded as
unaesthetic compared with the more compact solid tires, and there were
doubts about their reliability. Moreover, the safety bicycle was regarded
by some users as less macho than the earlier high-wheel "penny farthing."
But it was eventually shown that bicycles with pneumatic tires could go
faster than those with solid ones, even high-wheelers. Then, as Bijker,
Hughes, and Pinch explain it, the "social meaning" of the pneumatic tire
was redefined in terms of speed rather than reliability, and that made it
acceptable even to the boldest of sporting riders.[19]

Note again how the several social meanings of any technology differ
from the more personal meanings and perceptions with which this book
is mainly concerned. We saw that white goods had social meanings
regarding cleanliness, health and gender, but for some women, were part
of a restricting personal experience of narrowed-down horizons. By
contrast, the safety bicycle, when first introduced, was used by middle-
class women for outings with friends and as a reason for wearing less
heavy, less conventional clothing. It had a liberating social meaning and
enlarged horizons on a personal level also. At the same time, though,
men were seeking bicycles with a more macho image. But in terms of

personal experience, using a bicycle has other perceptions to offer, including the rhythmic, musical pleasure mentioned in Chapter 1.

Yet another example of the multiple social meanings of technology concerns light aircraft, and in particular, propeller-driven types carrying fewer than ten passengers. A great variety of designs is possible, and whereas most have tractive propellers, Lemonnier[20] identifies three designed in the 1940s and 1950s with propellers *behind* the wings working propulsively, and one Cessna design of 1961 with a tractive propeller at the front of the body and a propulsive one at the back. Other options that have scarcely ever been used include the canard design, with the stabilizer at the front rather than forming a tail. This design could have weight-saving advantages. In speculating why many of these design concepts have been held back, Lemonnier mentions some that pilots rejected, including the 1961 Cessna whose exceptional lateral stability made it seem too safe and easy to fly. Rather like the early safety bicycle, "the kind of safety given by this aircraft . . . did not fit the male image that a pilot has." Conventional aircraft shapes also seem to have social meaning for the general public denoting reliability and efficiency. It has become difficult for designers to depart from a conventional symbolism, and from widely accepted norms of what looks safe.

Yet another study of social meanings in technology refers to a French project that envisaged the rapid replacement of gasoline-fueled cars by electric vehicles linked to a radically altered pattern of transport use. Supporters of this plan saw the social meanings of the gasoline car in terms of an outdated industrial and urban scene that should be swept away. Critics of the project, however, believed that the existing type of car is too much embedded in modern society to be dispensed with in one revolutionary move. Its social meanings are too deep and wide-ranging for that to be possible.[21]

As with the pneumatic bicycle tire, this is an instance in which social meanings can alter, or be redefined, as the context changes. At some time in the future, the gasoline-fueled automobile may be given a more negative meaning by changes that make it seem out of date or an intolerable source of pollution. Moreover, the possibility of future shifts in social meaning seems more likely when we reflect on changes that have occurred in the past. During the 1920s, the saloon car stood for wealth and luxury.

But once it was manufactured in millions, the saloon inevitably lost that signification. Meanwhile, an actual 1920s model is today seen as a vintage car, with meanings involving some nostalgia for supposedly good old days.

Inevitably, advertisers and designers take up and manipulate these shifting meanings. Sports cars are given bullet shapes, or low crouching forms to represent a potential for springing into virile action, and one brand of gasoline was given similar attributes when advertised as "the tiger in your tank." By contrast, in the 1930s, a car with a domesticated look was shown being used by independent and fashionable young women under the slogan, "As dependable as an Austin."

Symbolism of various kinds—bullet shapes, glossy finishes, push-button controls, or computerized complexity—can be taken so far that artifacts may imply meanings without their functions being understood. On seeing a new household appliance, shoppers have been reported as exclaiming, "Gotta have it!" before they even know what it does. Similarly, people can become enthusiastic about new technology that looks like progress without really understanding its social meanings. Langdon Winner talks about "technological somnambulists wandering through an extended dream." He adds that it seems "all but impossible for the computer enthusiasts to examine critically the *ends* that might guide the world-shaking developments they anticipate."[22] Similarly, Pam Lin argues that educationists tend not to notice the wider social meanings of computers. They observe that girls in school often approach computing differently from boys, yet rarely carry their analysis of this to the point of recognizing that the computer is a "complex product of gender-differentiated, hi-tech workplaces."[23]

Public Responses and Political Meaning

The point that what technology signifies depends on the people interpreting its message is emphasized yet again by David Nye's historical study of public responses to major projects in the United States. He notes how people turned out in thousands during the nineteenth century for the inauguration of railroads, and how they have more recently attended such events as the fiftieth anniversary of the Golden Gate bridge, or the launch of a rocket. The American public, Nye says, "has repeatedly paid homage to railways, bridges, skyscrapers, factories, dams, airplanes and space vehicles." Enthusiasm of this kind is not related directly to the utility of the objects celebrated. Rather, this is something that "taps into fundamental hopes and fears. . . . It is essentially religious feeling.[24]

Elucidating further, Nye suggests that what people feel about technology on these occasions is comparable to how they react to a spectacular natural phenomenon such as might be described as "awesome" or "sublime." Appreciation of the sublime in technology also tends to be much more emphatic in America than elsewhere, he argues. Europeans did, of course, celebrate technology in the Crystal Palace and the Eiffel Tower. They admired prestige ocean liners and battleships, and later were proud of such full-throated creations as the first jet engine, patented in Britain in 1930 and first used in British military planes (1941) and civil airliners (1952). But nothing on the eastern side of the Atlantic could quite compare with flamboyant occasions in the United States when crowds converged on some great display of engineering or space technology.

One reason why Americans interpret technology in this way may be that the continent they occupy has sensitized them to the sublime by its vast scale and natural scenery. Nye finds some continuity between attitudes to the Grand Canyon and Niagara Falls and responses to great bridges or skyscrapers.

There was also a strong belief among United States citizens in "our manifest destiny to overspread the continent allotted by Providence." This view emerged strongly in debates about the annexation of Texas during the 1840s, and it seemed logical to celebrate technologies that helped bring it to fulfilment. As railroad building made it possible to extend settlement farther and farther west, there was a tendency to merge belief in "manifest destiny" with celebration of the railroad as the "technological sublime." Then in the twentieth century, technology appears to have taken over as the principal symbol of national destiny. Monumental bridges or skyscrapers and later space exploits became occasions for fostering "the sense of unity and the sense of future possibility that are essential ingredients in the achievement of political hegemony." It was as if the U.S. public had "come to depend on technology for periodic demonstrations of America's greatness."[25]

That is one kind of political meaning associated with technology. When we start to think of the larger systems within which many products are used—the railroad as a transport system, for example, rather than just the locomotive—we encounter another kind of political meaning. Indeed, Langdon Winner argues that technology has provided the body politic with a "new constitution," and as a U.S. citizen, he notes that this entails greater centralization of power than the founders of the republic ever envisioned. Winner is thinking here of oil companies and automobile manufacturers, with their concentrations of industrial power, as well as such networks of influence as the military-industrial complex. He argues that we should critically examine the characteristics of large-scale, sociotechnical systems, and suggests that this often reveals inconsistencies between changes in technology and the ideals of democracy and social justice. "All varieties of hardware and their corresponding forms of social life must be scrutinized to see whether they are friendly or unfriendly to the idea of a just society."[26]

Winner also argues that the political ideas implicit in large sociotechnical systems are often expressed by artifacts as clearly as by words. Such "artifact-ideas . . . tell us who we are, where we are situated in the social order, what is normal, what is possible, what is excluded." Among the many ideas "present in the structure of contemporary technological devices and systems," he identifies once again the assertion that "power is centralized," as well as a number of other political implications to do with social class, gender, and unequal distribution of wealth.[27]

In much discussion of this kind we need to be clear about how much of the social and political meaning of technology is encoded in artifacts, and how much is expressed by the larger systems. The design of clothes washers as compared with video recorders may reflect assumptions about the competences and interests of women as distinct from those of men. But in most industrialized countries, it is the system of electricity generation, or of manufacturing consumer goods, or of television broadcasting that reflects centralized power most strongly. In other words, the details of an artifact's design may have social meanings related to gender, whereas the system of which it is part has political meaning reflecting the concentration of industrial power at the expense of democratic values.

According to Langdon Winner, no successful industrial society has resolved the contradiction between democratic and technological values, and he sees this as a significant failure of modern civilization. It arises because we neglect to keep questions concerning democracy high enough up the agenda. Indeed, we allow them to get lost behind assertions that technology is politically neutral. Decisions about adopting new technology are then based on criteria of economy and efficiency, not justice and freedom.[28] Crucial questions about the cultural environment such technology will be used to create, and who will control it, are never adequately answered. Thus one may observe public taste manipulated by owners of the electronic media; community activities undermined by promoters of mass entertainment; and employees' rate of work monitored by computers (as was already happening to thousands of Americans in the early 1990s). Quoting evidence for that, Winner notices how "many people in freedom-loving countries like the United States seem eager to embrace repressive models of social integration," provided that the repression is mediated by glamorous and sophisticated technology, as with electronic monitoring of work.[29]

This chapter is not the place to attempt a resolution of these contradictions. The point to make here is simply that tools and machines are not merely neutral aids to production. Nor can engineers and scientists claim that they deal only with technical problems in a value-free and neutral manner. As commentators frequently point out, the characteristic pattern of research in the twentieth century has been that a connection with industrial or military power allows engineers and scientists "to dream in an expansive fashion," and the huge resources commanded by that power "brings their dreams to life."[30]

Multilayered Meaning

One of the aims of this book is to suggest a framework for discussing experience of technology, and what has emerged here is the contrast between individual experience and the shared experience of wider groups. A cook experiences the textures and aromas of foodstuffs being mixed or heated, and that is the individual experience that he or she uses to make judgments, again as an individual, about how the cooking process is going. But the cook is also aware of the shared, social meaning of the meal for the people who will eat it. The designer of a locomotive or an airplane may personally experience considerable satisfaction as the concept takes shape on the drawing board or computer screen, and his or her visual responses may inform some technical judgments. But the designer must also consider whether the finished machine will have an appropriate social meaning for the passengers it will serve—whether it will look safe and reliable, for example.

It is possible to think of these various meanings of technology in terms of a hierarchy of levels, with individual experience on the most private level and the experience of the consumers of cooked food, the train passengers, and other groups as a more public, social level. Beyond that, the political meanings of technology discussed by Langdon Winner and others are on the level with the widest ramifications of all.

Questions concerning these different levels of meaning, individual, social, and political, also arise among art critics. Some traditionalists interpret pictures in terms of the artist's skill in handling the paint, or in arranging a composition, and may see these things as a source of elevating individual experience. But those who wrote like this in the 1990s were liable to find themselves confronted by "an aggressive phalanx of critics and theorists who assert that all art, like the rest of culture, is politicized to the core." Exchanges between schools of thought were said to "descend rapidly to distracting invective."[31]

However, both viewpoints are valid, because we need to take account of individual experience as well as political meaning. If paintings are effective as social or political statements, this is partly because the artist has used individual experience to refine his or her skill and judgment, and not just because of the artist's political circumstances and commitments. There is then a convergence of individual and sociopolitical meaning in the work of producing the painting.

Similarly, we may analyze the significance and meaning of computers in ways that demonstrate convergence of the inventive experience of creative individuals and the political or economic conditions affecting their work. One study that portrays such a convergence was produced by a "radical science collective" with a strong interest in emphasizing political meanings in computers, yet with a keen awareness of the creative experience of many of the individuals involved. The collective was critical of commentators who write about computers in education only in terms of the individual, personal responses of students, yet recognized that the personal experience can be vivid and real. They discussed research on artificial intelligence in relation to "an embarrassing entanglement with power," but recognized also the personal excitement and "joy" of creative, exploratory research in this field.[32]

Philosopher Carl Mitcham has offered a way of understanding situations such as these by noting that several meanings are denoted by the word "technology" itself, and suggesting that these multiple meanings can fruitfully coexist. Defined in one way, computer technology does indeed mean entanglement with power, whereas defined in another way, it is all existential joy. Allowing the two meanings to be simultaneously applicable implies a multilevel view of the subject and calls for a "pluralistic philosophy" in relation to technology.[33] Part of this pluralist view must clearly be the responses of individuals, but we also need the insights into social meaning of technology offered by semiotics, social constructivism, and studies of the technological sublime. Beyond that again, we need political analysis of technological systems and artifact-ideas.

The theme of this book, however, is the personal level of experience rather than any of the other levels, and so far, the focus has been somewhat utilitarian. We have discussed personal responses and individual experience in relation to inventors tinkering with cathode-ray tubes, engineers enjoying sweetly running machines, and cooks being aware of textures and scents. All these forms of experience have a role in routine practical judgment. They are essential to the practical skill of the craft worker, as we have seen in the two preceding chapters. What also needs recognition, though, is that a great deal of human behavior in these contexts is essentially playful and that other impulses, describable as "aspiration," enter into experience of technology.

Play and Aspiration

According to those who study animal behavior, most of the higher animals may be observed to play while they are young. Cats, otters, and chimpanzees are examples often described. But whereas most animals tend to play less as they grow older, humans do not so much lose the urge to play and explore, but rather they systematize it and organize it better. They superimpose "play rules" onto the more serious preoccupations of adult life.[34]

Several authors make it clear that they see scientific research in this light, one noting a "smooth transition from inquisitive childhood play to the life-work of a scientist," and another, himself a scientist, agreeing that: "All science, all human thought, is a form of play."[35] By comparison, then, technology, or at least its hardware, might be regarded as beginning with toys.[36] Thus George Basalla has shown how playfulness is involved in the process of invention in just about every situation one can imagine. His starting point is the observation that inventions are too numerous and diverse to be accounted for by necessity and utility alone. Necessity is *not* the mother of invention, he claims. In some cultures, at least, wheeled toys existed before any practical use was found for the wheel. In Europe and America in the 1880s and 1990s, the automobile was a toy, a "play-thing for those who could afford to buy one."[37]

This point is often made, but its significance is not fully brought out because conventional explanations of the development of technology portray only the responses of innovators to economic conditions, or speak of the social construction of useful artifacts. Basalla argues that psychological factors also need to be brought into the equation, and that we need to understand human nature as essentially playful. The species should be classified as *Homo ludens*. Playfulness for Basalla is exhibited in puzzle solving, meeting challenges, competitiveness, and "technological dreams," sometimes expressed as science fiction as well as in real science or invention. There was always in human culture "a technological imagination that took delight in itself." And whereas some inventors thought they could make money, others pursued novelty "for the psychic rewards it brought."[38]

Some inventors, of course, made fortunes just by developing toys, and some toys had a stimulating role in technology. An example was the toy construction system famous for half a century as Meccano. It was patented in 1901 by Frank Hornby, manufacturer of model trains, whose toy factories had made him a millionaire by 1915. His concept was a system of metal plates, angles, and strips that could be bolted together to make a great variety of small machines, gadgets, and structures. The holes through which bolts were inserted in plates and strips or within which axles could turn were about 3 mm (one-tenth inch) in diameter and were spaced at regular 13 mm (half-inch) intervals. There was subtlety in these dimensions because, if the holes had been bigger or more closely spaced, the metal strips in which they were cut would have been too weak and flexible. But had the holes been smaller, the axles that had to turn in them would themselves have been too thin.[39]

Meccano was very much a product of the time when bridges and ships were built by riveting metal plates and angles through regularly spaced holes. It lost popularity in the succeeding age of plastics (which spawned its own construction toys), and then with the advent of computer games. Older forms of play have persisted more strongly. Iona Opie, a folklorist who has recorded toys and games, notes the continuity of some of them from the time of the ancient Egyptians and of the Greeks, who had hoops, tops, balls, yo-yos and marbles.[40] A psychologist adds that boys and girls tend to play differently from a young age, and that boys' play develops better visual-spatial abilities. This, combined with a tendency to be more playful in adult life, may partly account for a greater male involvement in some kinds of technology and invention. Play with construction sets such as Meccano also contributes to the development of visual-spatial skills. Research finds girls keen to use such toys when given the chance, but teachers and parents, until recently, more often provided toys that would perpetuate gender stereotypes.[41]

Yet another form of play of interest to students of animal behavior is collecting, whether of stamps, coins, used phone cards, or fossils. The pleasure people get from collecting is related to the satisfaction of classifying objects. Biologist Nicholas Humphrey remarks that: "Nature makes behaviour pleasant so that animals will do it." Because animals need to recognize the safe and edible as compared to the dangerous and toxic, the art of classifying objects into these categories evolved as something pleasurable. "Indeed, evolutionary pressure to classify may have been as strong as the pressure that made eating and sex so efficient and enjoyable."[42]

Authors who argue this way also point out that although curiosity, exploration, and learning have survival value for animals, such aptitudes occasionally develop beyond the limits of utility, as when magpies collect objects at random in their nests. In the human, though, the urge to collect and classify has become extraordinarily prominent, and in it, some writers such as Humphrey see the origins of scientific research. Many other things that happen during childhood play have a bearing on technological invention. Drawing, building with toy bricks, learning one's own abilities and how to extend them by using objects as tools are all involved. It would not be surprising, then, if some of the meanings we find in technology are related to impulses originating in play that have become serious adult preoccupations.

There may be other reasons, though, for the interests and commitments that some people display in their attitude to technology. At the start of a book that otherwise has little to say about such things, David Noble comments that the practice of technology includes "a subjective element which drives it." One might wonder if this is a play impulse, but Noble makes it sound much more serious as he quickly moves his argument onto a political level. Thus he identifies the subjective element with the personal energy of "the most powerful and forceful people in society in struggle with others."[43]

That characterization suggests an inner personal drive unconnected to rationally defined objectives. Some commentators who recognize the same sort of impulse describe it as a "will-to-power" or see it as part of the "technological imperative." Others (speaking of architects rather than engineers) say that "like all men of action, they may not understand the meaning of what they do."[44]

When pressed for an explanation of what lies behind these strong but apparently blind impulses, scientists speak of curiosity about nature, engineers resort to statements about the "technically sweet," and architects claim that their buildings have "grown from a love of materials." If the present book were misinterpreted as merely exalting the pleasures of working with materials, as the previous chapter partly did, it would contribute to these tendencies. My intention, rather, is to warn that those concerned only with political analysis often ignore the heady brew that "love of materials" and similar expressions can hide. It is necessary and right that engineers, scientists, and architects enjoy the materials they work with and the sweet running of the machines they build, but it is always dangerous when these things become ends in themselves.

Attempts to account for these impulses and imperatives today find some clues in studies of play behavior, but at one time it was more usual to associate them with the male sex drive. Thus nineteenth-century novelists such as Elizabeth Gaskell may have thought that way when they portrayed the industrialists of their time as thrusting, energetic, single-minded men. Critics of a certain generation have been keen to point out every possible image used by Gaskell and others that might connect technology with male sexuality of a "dominating and destructive kind." In a caricature of such interpretations, David Lodge mentions "tall chimneys thrusting into the sky" and "buildings shaking with the rhythmic pounding of mighty engines."[45]

Similar interpretations are sometimes offered for the church buildings, spires, and elaborate public clocks of medieval Europe. These artifacts clearly represented the civic and political power of ecclesiastical authority. They certainly sometimes suggest sexual imagery. Yet they also strongly express what are often described as higher aspirations. Spires can be seen as arms reaching upward toward sources of heavenly inspiration: hands seeking some grasp on the transcendent. Like all important symbols, they had multiple meanings. What they signified again depended on the knowledge and values of the observer. It would be wrong to exclude sexual themes, but in considering William Golding's remarkable fictional interpretation of a spire, with its phallic shape, one should notice that Golding actually shows how different people—different receivers of the building's message—respond to other kinds of meaning as well. One of his most vivid lines, indeed, refers to a man for whom the spire was a "stone diagram of prayer."[46] That line is quoted by Samuel Florman in his essay on existential experience in engineering, and Florman claims that every major work of engineering has something of the cathedral in it: a similar sense of reaching upward and beyond.[47]

In a comparable way, Koestler notes that a spire seems to "soar" upward, making our thoughts soar with it, and the spires of Chartres have been interpreted by others as signifying "aspiration, a rising skyward, for medieval imagery is full of ideas of ascent." Spires were "powerful symbols of endeavour, of the upward flight of the human spirit." Expressing much the same feeling, "the rocket, tall and slender, awaiting a thrust away into outer space, is . . . part of twentieth-century iconography."[48]

Conclusion

Play is often seen as a lighthearted activity that lacks serious purpose. The building of towers and spires, by contrast, was a deadly serious and dangerous occupation that one feels ought to have had very clear purposes. Yet lacking other explanations of what these purposes might be, one talks about aspiration and diagrams of prayer. Or David Noble suggests that a subjective element drives major projects in technology, and others speak of impulses arising from masculinity or a more abstract "will-to-power."[49]

We are clearly dealing with a human motivation that people find difficult to put into words, so it may be helpful to recall the discussion of music in Chapter 1. There it was argued that music can often be strongly purposive in its energy, flow, and organization, but without having any defined goal or purpose at all. This is a paradox to which Immanuel Kant alluded when he asserted that "purposiveness can be without purpose." Here, I am guided by Körner's explanation of what Kant meant, namely that when we find a plant, an animal, a piece of music, or a medieval spire whose parts are "so intimately related and so harmoniously fitted together" that they imply coordinated development, then we interpret them as being purposive in a generalized way, even if we cannot identify any explicit or defined purpose.[50]

Alternatively, we might say that the only purpose of a piece of music is to be itself. And the purpose of a cat, in Kantian language, is "the cat's having to be." Such odd statements are needed to approach the paradox of "purposiveness without purpose," and we commonly encounter this sort of purposiveness, I suggest, in human play and creativity, and in animal behavior also. The difference between humans and animals is that humans have language and conscious thought, and can attach the purposiveness they feel in their lives to specific goals and defined purposes. When we talk about the social meanings of technology, as we have in this chapter, and when we talk about the individual experience of a cook trying to get a recipe right, or an inventor trying to get a machine to work, we are talking about specific purposes that can be formulated in words if we pause to analyze matters. But when we talk about will-to-power, or diagrams of prayer, or the upward flight of the human spirit, we are talking about generalized purposiveness and undefined aspiration. As in the purposive behavior of animals or young children, this refers to impulses which have a co-ordinated direction, but are not linked to consciously formulated goals. Like play, such impulses have probably always had a role in technology and are a familiar part of most people's everyday experience. Even ideas such as "progress" seem to refer partly to purposiveness of this kind, for so much of what is claimed as progress is merely coordinated change, lacking clear goals.

Purposiveness without purpose is especially characteristic of the lives of individuals whose technological creativity seems driven and obsessed. The influence of such people, when they become part of a research institution or weapons laboratory, can easily emerge on a political level, and urgent but unclear imperatives can infect entire pojects. Then commentators begin to write of "runaway technology," or complain of innovation that seems blind to social purposes.

This has a bearing on the initial question of a framework for discussing individual experience of technology. The framework used in this chapter has so far been based on a hierarchy of different levels of experience, from the private and individual to the public and political. That idea is used reluctantly, as explained in the introduction, and I have no intention of suggesting that the personal is more fundamental than the political, or vice versa. Now, though, it seems important to include different experiences of purposiveness within this framework.

For example, the private visual and tactile experiences of cooks, craft workers, engineers, and others in relation to specific practical tasks are related to practical judgments and, through those judgments, to defined purposes. So in cooking, private, personal experience related to tactile qualities, taste, smell, sight, and the cook's aesthetic responses to these experiences all influence judgments, related to the purpose of, say, baking a cake. But this specific purpose involving an individual's experience is usually related to wider, social-level purposes that may focus on the welfare of one's own family, or the social meanings of a shared meal with friends, or the financial goal of keeping within a budget.

By contrast, there is the experience of those who get wrapped up in solving puzzles or building ever more elaborate machines, or who seek esoteric knowledge, and who feel that these things are ends in themselves. Their work often seems embued with dedication and purposiveness, but without any defineable social purpose. It involves many subsidiary tasks with defined goals, but the source of the individual's energy and dedication is often an undefined sense of aspiration, such as may best be paralleled by the purposiveness of a great piece of music.

Once again, we invoke music, not just as a casual analogy, but with the conviction that this is the best metaphor available for experience of technology. First, music exemplifies purposiveness in the Kantian sense of co-ordinating and organizing parts to create wholes. This co-ordinating process is central to design, and as such is often harnessed to a defined purpose; but it is also capable of becoming an end in itself. It is then expressive of aspiration, drive, progress, prayer, the upward urge, and all the other strange words that occur in technological rhetoric. Second, though, music provides metaphors for understanding nature, as suggested in Chapters 1 and 2, and if technology is a major part of the human response to nature, it needs to be informed by this second aspect of musical sensitivity and its insights into purposiveness.

All this is apart from the role of rhythm and sound as a complement to the visual and tactile experiences of technology. On this level, musical experience can enter into specific judgments in the same way as visual experience.

Historians and sociologists of technology, if they have the slightest positivist inclinations, are reluctant to acknowledge that this kind of experience can be important, because it is not susceptible to objective, scholarly study. There is certainly scope for skeptical analysis of rhetoric about progress and aspiration. But in other respects, one only has to be a human being and participate in music or in some practical task involving sight, touch, and smell (such as cooking or woodwork), and one knows that individual experience is centrally important.

Figure 3
Creativity in the context of the family, with father a harlequin and musician
(drawing by John Nellist)

II

Contexts of Technology: Nature, People, and Conflict

5

The Sense of Place

Environmental Perspectives

In Africa there is a landscape of rock surfaces and boulders, scrub and grass, whose hills offer long views, and where an apparently tiny structure is a rectangular stone slab on a low plinth. It is the grave of Cecil Rhodes, founder of an alien colony in this land. Cut into the solid rock of the Matopo Hills, it is seen by some as an intrusion upon a traditional sacred site. The rectangular geometrical shape is expressive of a technological ability to master even the hardest rock, and it clashes with nature's freer shapes.

The site illustrates two kinds of meaning that we tend to find in landscape: a sense of nature as something beyond us, before which we stand in awe, but also a feeling of being invited to leave our mark. A famous character in Chinese literature named Monkey, once bounded to the very edge of the universe, where he found five great columns. He should have felt wonder and respect, but his impulse instead was to leave his mark, so he urinated on one of the columns—on one of the fingers of God.[1]

Another significant landscape lies among the tablelands and mountains of New Mexico, where D. H. Lawrence (in 1923–24) and other literary figures once took refuge. They sensed this to be a place apart from the busy, urbanized, industrializing world, and some were attracted also by the fact that this was still a sacred world for the Navajo Indians, whose country it was. The irony is that, during the 1940s, this place of escape became the setting for the most ambitious effort hitherto for wresting nature's ultimate forces from her control, because then the Los Alamos laboratory was established there, and a community of atomic scientists took up residence.

Commenting on the stark irony of this situation, Debra Rosenthal has remarked that efforts "to regain the lost world of . . . a pretechnological paradise are easily ridiculed as naive and vulgar romanticism. In western culture, the vision of life in harmony with nature was superseded long before . . . Lawrence took refuge in the raw mountains of northern New Mexico." But still "somehow the dream of harmony persists, even in the shadow of the bomb."[2]

Today, with the Cold War over, we perhaps feel less threatened by the bomb (although it still is a threat). But many people feel far more threatened by the prospect of environmental catastrophe, and are probably right to do so. Indeed, this threat may be considerably more serious than most commentators admit. I often think that if people realized what is really at issue, they would immediately abandon their cars, switch off their refrigerators, and eat only organically grown vegetables with no meat or fish. Or they would boycott the energy businesses whose lobbying has (in 1997) prevented effective agreement to reduce greenhouse gas emissions. Why, then, is this chapter not dedicated to promoting a fridge-free, vegetarian lifestyle (a subject on which I am peculiarly well-qualified to write)? Or, to be more realistic, why is it not concerned with political analysis of the business interests that are so busily encouraging pollution and subverting research on the subject? Why then do I start by talking about landscapes in New Mexico and Zimbabwe, and quote from a Chinese fictional work?

One reason, of course, is that as in the rest of this book, I believe that political analysis, which is absolutely essential, needs to be complemented by an understanding of how individuals experience the environment. In this chapter, that will mean especially how they experience landscape, whether in New Mexico or in their own home area, and how they feel about marking the landscape, either as Monkey did it, or by constructing dams and irrigation systems, or bridges and highways.

A second reason why I am not tackling the worldwide environmental crisis head on is that much of the usual discussion relies on forecasts of population growth, resource use, pollution emissions, deforestation, extinction of species, and climate change. Although these topics represent enormously serious issues, they are frequently misrepresented. Some oft-quoted facts about deforestation in Africa are simply wrong, for example, and invite damaging solutions to imaginary problems.[3] Predictions about resource utilization and depletion, beginning with the work of nineteenth-century economists,[4] and continuing with modern systems theorists,[5] have usually proved to be overpessimistic. Yet when the subject is properly presented, we find that there is an inescapable limit to human use of the environment.[6] Moreover, even if we cannot predict what may occur, history records earlier environmental catastrophes in which the human population was drastically reduced, and we can surely learn from them.

But although this second part of the book again offers only a discussion of individual experience, rather than analysis of the world situation, it attempts a wider perspective than Part 1. It tackles larger themes and different points of view, being less oriented to the experience of engineers, designers, or users of equipment. Its approach is clarified if I refer again to the point made in the Introduction about my experience in 1951, when I was introduced to technology as a subject with two contrasting themes. On the one hand, there was pleasure and excitement in making things (or seeing and understanding how they were made), with enjoyment of their architectural and musical characteristics. On the other hand, there was the impulse to make life better for people: to ensure that material needs were more adequately met, to relieve suffering, and to enrich quality of life.

For many engineers and scientists, these two themes complement one another. In their work, enthusiasm for the aesthetic purposiveness of technology is linked to socially useful applications. However, my own perceptions during the half century since 1951 is that these two kinds of purpose in technology have been pulling apart. On the one hand, I am as sensible as ever of the thrill of discovery and creativity, and of the musicality, aesthetic achievement, and craft skill to be found in the practice of technology. I have used Part 1 of this book to celebrate these things, to assert their value, and explain their importance, even while warning against their seductiveness.

On the other hand, it has become clear that all this wonderfully transcendent purposiveness is often out of step with social purposes that need to be addressed. When engineers and scientists turn from talk of discovery and creativity, which always commands respect, and instead make claims about how society will benefit and how life will change for everyone, feelings of skepticism, cynicism and even disgust at the complacency of such claims overwhelm my initial curiosity and interest. Similar promises were made in 1951, specifically about the benefits of nuclear power, about improvements in everybody's quality of life, and about how science would eliminate malnutrition and starvation throughout the world. Today's promises about how agricultural science will feed growing populations ring equally hollow in a world where malnutrition seems no less prevalent, and when agro-industry is protected by "food disparagement laws" that limit open discussion.[7]

In later chapters, we may gain further insight into this paradox of what is valuable in science and technology, and what seems to betray its social meaning. Here it suffices to consider what might be meant by saying that quality of life for some people (myself included) has deteriorated since 1951. This statement, of course, is contrary to the statistics for standards of living in industrialized countries, which show steady improvement. However, those are statistics based on data for gross national product (GNP) per capita, and critics point to incidents such as the wreck of an oil tanker to show the inadequacy of GNP in this context. After such an incident, the millions of dollars spent on cleaning up beaches have put extra wages in the pockets of the workers employed and led to extra sales of detergents and other materials used, all of which add to GNP. Thus an event that diminishes quality of life for many people is recorded in the statistics as an increase in standard of living, because only the positive impact on wages and sales is measured; the negative impacts are widespread, hard to measure, and ignored.

Given the unsatisfactory nature of statistics based on GNP, we might refer to figures for expectation of life and educational achievement, most of which have also consistently improved since 1951. Or these data can be combined with GNP data to produce a "human development index," as is done in an annual United Nations report that tends to show how Canada and Japan have very good quality of life, with the United States and Britain some way behind. However, all these measures depend on what one most values, and for many people, quality of life ought to be related also to stress (or lack of it), hours available to spend with their children, and various aspects of the environment, such as noise levels (which increase remorselessly). Thus when it is asserted that quality of life in the United States (for example) has declined by some specific percentage during a decade when per capita consumption steadily increased, it is hard to believe that the precise figures have any meaning.[8]

Although it would be nice to have something unambiguous to measure, quality of life is ultimately about what people experience, or how they respond to it, and whether the experience and response together enhance their well-being. Once again, therefore, it is necessary to consider individual experience, as Langdon Winner does, for example, in describing the part of California where he grew up, between Los Angeles and San Francisco. He mentions new highways invading the surroundings of some communities, and recalls that "in a few short years the town witnessed the coming of freeways, jet airplanes, television . . . food additives, plastics. . . . The shape of the house and the activities of the family were refashioned to accommodate the arrival of all kinds of electronic gadgets."[9]

Whereas people in general accepted most of this innovation and change without question, taking it all for improvement and progress, Winner balances gains and losses more critically. The one clear gain he records was the coming of the Salk vaccine. The losses include a reduced availability of fresh food and its replacement by a less enjoyable, more heavily processed diet. But much of what he says concerns the environment, though with a rather specific emphasis. Where he could have said much about the elimination of plant species or instances of pollution, what comes over more prominently is how the home has been altered, how attractive buildings have been replaced by characterless ones, and in general, how the surroundings of the town have been filled up with rather anonymous constructions.

Today social scientists and philosophers discuss people's responses to environmental change, and they would say of Winner's experience that his sense of place has been offended. Buildings and countryside that gave his area identity have been replaced, and even the home has been reshaped, by implication making it less like home.

Valuing "Place" and Relating to Nature

In a study of what might be meant by "sense of place," philosopher Jane Howarth notes that there are at least two ways of assessing the value of a landscape, both of which may be relevant when it is to be changed by a planned development.[10] First, we can catalogue rare species, assess biodiversity, and note habitats vulnerable to disturbance or pollution. We may even (dubiously) attempt a cost-benefit analysis to assess gains and losses likely to arise from the development. These efforts reflect a scientific approach in which we regard ourselves as separate from the landscape and generally detached. We attempt to assess all the issues objectively. When we adopt this point of view and think of the people who enjoy the landscape, we tend to regard them as using it as a playground or taking pleasure in a spectacle.

Second, though, if we live in and like a place, none of this scientific analysis expresses quite what we feel about it, and we may think that we ought to assess the locality from this other point of view also. That means exploring what we mean by "sense of place." When I write reports on planned developments in my own area, I discuss the archaeology and botany of the landscape as scientifically as I possibly can, but I also try to explain in what way the landscape helps to identify my community, Addingham, and to define its location as a significant place. Speaking generally, Jane Howarth suggests that what we often feel about the place where we live is "attachment" of a kind that "goes very deep, is of significance in the life of the individual . . . (and) is an important part of being human. It is comparable with one's attachment to one's closest friends."[11]

Such feelings may also include a sense of attachment to "nature as nature," rather than to nature as statistics or Latin names for species. When thinking about a place in its totality, we do not separate ourselves as subjects from the place as object, but consider ourselves as part of the place. Similarly with nature, so that some philosophers have said that we are then taking a *participatory* rather than a *detached* view of life.[12]

These are very strong statements, and some may doubt whether anybody in the modern world really feels like this about Nature (now often with a capital "N"). Or if they do, aren't they being excessively sentimental and certainly prescientific? It is of interest, then, that some exceptional scientists seem to have depended on a participatory attitude to Nature of precisely this kind. Thus Barbara McClintock's researches on maize (corn), cited in a previous chapter, were motivated by a "feeling for the organism," and Edward Wilson, another modern biologist, remarks that the best of science "doesn't consist of mathematical models and experiments. It springs fresh from a more primitive mode of thought."[13]

In the nineteenth century, Michael Faraday's extraordinarily fruitful researches on electricity were "a face-to-face, heart-to-heart inspection of things." His diary is a record of intimate dialogue with Nature, posing questions and waiting attentively for answers. The "emotional basis of Faraday's science" was humility and a sense of wonder and joy in the natural world.[14]

Such responses to Nature could often be linked to specific places, as when Joseph Banks, almost a century before Faraday, decided to become a botanist after finding himself alone in a country lane surrounded by wild flowers.[15] Determined rationalists, of course, have no sympathy for these attitudes. For them, Sir Isaac Newton's wonderfully logical account of the motions of the planets once served as a powerful example of how a rational, mathematical understanding of phenomena was possible, free of emotional "enchantments." But now we know that Newton had sympathies for certain alchemical ideas regarding nature that "he dared not publish," even though they had contributed to his concept of gravitation. In these words Morris Berman sees the "disenchanted," rationalist view of nature as founded on self-censorship and "buttoned up" feeling— on separating oneself from nature (now with a small "n") and abstracting from it only those things that can be measured and calculated.[16]

Despite the many insights and material benefits that come from looking at the world in a detached way, many people still feel that there should be acceptable ways of acknowledging their own responses to sun and sky, mountains and oceans, and the burst of new life at every springtime. Forests and seashores are still places to which we can feel drawn. The ocean has its own "strange power . . . which fills our language with its metaphors," as mountains still seem to have "presence."[17] One such is Beamsley Beacon, a hill close to where I live that constantly draws one's eye. In the grander landscapes of North America, a correspondent reports, the Rockies also have presence to which people in Calgary react strongly, often with respect and exhilaration, but sometimes with a sense of claustrophobia. It is easy to understand why people untouched by disenchanted science sometimes identified hills with spirits. Conversely, it is also easy to see why the rational men of the eighteenth century turned deliberately away, like the self-censoring Newton, or like the travelers of this period who pulled blinds across the windows of their coach to avoid having to contemplate the mountains of the English Lake District.

Among thinkers of the Age of Reason who were prepared to look at mountains, some analyzed the feelings that arose by saying that the mountains were "sublime" in the sense of being awesome and thrilling, whereas many other aspects of nature were "beautiful" in a less threatening way. This distinction was made by people who had lost the old visceral sense of Nature as alive and organic, but who still felt an emotional response. It was a distinction made by philosophers, including Burke and Kant, but the fact that they wished to speak of the sublime implies that they wanted to recognize their emotional reactions to nature rather than dismissing them as unimportant. And it seems futile to deny that there has been some sort of appreciation of landscape and nature in nearly all civilizations and cultures. Indeed, certain responses to nature seem inextricably linked to feelings of attachment to territory, to the sense of place, therefore. They are not just the product of the romantic movement.

However, we need to acknowledge that the romantic view can be seriously one-sided. Think of the painter or poet who saw the countryside only on fine summer days and had no experience of what it was like to work in the fields in all weathers. Think also of today's hikers who find relief from the pressures of urban living in the quiet of the Welsh hills unaware that farmers in the area are under greater economic pressure than most city dwellers, experiencing more depression (more often leading to suicide) because of the isolating loneliness of a landscape that makes living so hard.

A century ago, in the fen country around Ely in Cambridgeshire, many acres of land would often flood in winter, and then "the little fen villages seated upon their small hills" stood up out of the water "like castle-crowned islets in Swiss lakes." Some people went skating when the floodwater froze, and there were days of "picturesque beauty" as in paintings by Dutch masters. But for those who lived in the villages, these floods could mean tragedy. Rarely was life "so starkly grim."[18]

Poverty was extreme in many other parts of the English countryside during the nineteenth century. It therefore gives pause for thought that those who had experienced the hardships of rural life and came to write about them could still appreciate the beauty of their surroundings. An outstanding example is John Clare, the "peasant poet" of Northamptonshire. His editors comment that much of what he wrote could have become merely sentimental in the hands of a more conventional writer, but what made the difference was that Clare "knew village life from the inside." He referred to the regular periods of unemployment that were part of the farming year as "leisure's hungry holiday," and knew all the agonies resulting from the enclosure movement. At the same time, his observations "of flowers and bird life are those of the finest naturalist in all English poetry."[19]

The paradox of natural beauty in a rural scene full of oppression is more explicit still in Flora Thompson's description of harvest in the English Midlands. Having grown up in a laborer's cottage, she remembered "night scents of wheat-straw and flowers . . . and the sky . . . fleeced with pink clouds. For a few days . . . the fields stood 'ripe unto harvest.' It was the one perfect period in the hamlet year." The work of harvest, too, was enjoyed when, "in the cool dusk of an August evening, the last load was brought in." But then comes the sharp stab of reality, as Thompson remarks that it did not do to look below the surface and notice the starvation wages. Describing the harvest celebration, she remarks: "The joy and pleasure of the labourers in their task well done was pathetic, considering their very small share in the gain. But . . . they still loved the soil and rejoiced in bringing forth the fruits of the soil, and harvest home put the crown on their year's work."[20]

So in this inquiry into the meanings people find—or construct—in landscape and in work on the land, it may be worth reaching back to an earlier period, beyond the contradictions of nineteenth-century romanticism. In medieval poetry, for example, one finds a powerful feeling for Nature in the world of Hildegard of Bingen and Francis of Assisi, as well as in thirteenth-century sculpture portraying leaves, fruit, and flowers. One can find it also in the way Thomas Traherne, during the seventeenth century, wrote about his sense of identity with Nature: "Your Enjoyment of the World is never right, till every Morning you awake . . . and look

upon the Skies and the Earth and the Air as Celestial Joys . . . till the
Sea it self floweth in your Veins, till you are Clothed with the Heavens,
and Crowned with the Stars."[21]

This represents what was referred to earlier as a participatory consciousness—a sense of being involved in nature— and we contrasted that
state of mind with the detached consciousness that has been associated
with the growth of science over the last three centuries. Intermediate
between the two is the sense of wonder at and longing for harmony with
nature expressed by some modern scientists,[22] as well as poets and
painters.

Disregarding this intermediate position, there is a contrast to be drawn,
then, between two ways of looking at nature. On the one hand, there is
participatory experience of the vividness and purposiveness of everything
in the world, and on the other there is the more detached outlook within
which all such talk is fantasy. Some people would say that to acknowledge feelings of any kind can only get in the way of a proper scientific
approach. Individuals who take this view prefer their science to be
presented as "the experience of no one." Their thinking tends to be
object-centered (as defined in Chapter 2), and they seek to avoid working
in an involved, participatory way, which they think would lead to bias.

Participatory Technologies

Many of the traditional craft technologies discussed in Chapters 2 and
3, including wheelwrights' work, pottery, and many kinds of metalwork,
were practiced in a participatory way, with the individual worker feeling
a strong personal involvement with materials, and making full use of the
vital immediacy of sight, touch, and other senses. The skills of the soil
scientist have also been mentioned in drawing a comparison between
detached classroom experience and moments of "participatory" insight,
when soil was actually dug up and felt between the fingers.

Of course, prescientific peoples in all parts of the world required
knowledge and skill related to the landscapes in which they lived, and
inevitably, this knowledge was at first of a participatory kind. It was the
knowledge needed for hunting, gathering, or growing food and for obtaining other necessities: materials for making shelters, fibres for ropes

and clothing, herbs for medicinal use. And it was knowledge that could be gained only by experience of the most practical, involved kind.

It is often assumed that early human populations could exploit their local landscapes to obtain food, fuel and shelter, without disrupting the environment, but this is another romantic illusion resulting from modern mythmaking. Our tendency to assumed that all "primitive" peoples lived in harmony with nature is a reflection of what we would like nature to mean for us. No human groups ever had a painless way of fitting into their environment. Some groups, indeed, devastated large areas, or employed destructive methods of hunting (for example, driving herds of buffalo over cliffs).[23] Attitudes and skills capable of correcting such damaging activity were learned only slowly.

During the thousands of years in which humans have lived in Australia, many of the larger marsupials were forced into extinction and the landscape was modified by systematic burning of vegetation. The philosophy of harmony with nature developed by Aborigines in more recent centuries is the result of a painful earlier process of learning to curb destructive tendencies and to live in a way that the landscape could accommodate. Not all peoples achieve this, and it is not true that early hunter-gatherers were instinctive conservationists. Those who survived into recent times were able to survive precisely because they managed to learn restraint, often by developing mythologies that encouraged a "reverential attitude to the creatures they kill, and to nature as a whole."[24] Any surviving descendents of twentieth-century civilization will, in the long run, be those who similarly evolve an attitude of restraint.

Another kind of technology related to landscape (and seascape) in which non-Western peoples were often highly skilled was navigation in trackless deserts, in snowy wastes, or at sea. On the Pacific Ocean, for example, people could travel by canoe from one island to another, undertaking voyages lasting several days out of sight of land. Their navigation techniques depended on integrating several kinds of sense experience relating to winds, waves, seabirds, the smell of distant land, the apparent color of water over reefs, and the sun and stars. Swell patterns in particular could provide many clues to the location of islands, and were recorded by means of "stick maps" formed by lashing slender sticks together in complex geometric patterns.

Although Polynesian seafarers have lost many of these skills through contact with the West, people of the Caroline Islands have continued to practice traditional navigation and have intrigued and puzzled mathematicians with their skills. Visual thinking of a high order is involved in distance estimates, as also in using astronomical knowledge to represent a conceptual "star compass." But serious mistakes are rare, and the smells, swell patterns, and bird life associated with the destination island are usually observed at the expected point in the voyage.[25]

Travelers on land also used the stars for navigation, especially across the deserts of the Middle East, but landscape features more often provided means of establishing position and direction. Thus the Inuit people of the Arctic can undertake long journeys in apparently featureless tundra and ice fields without getting lost because they pay close attention to snow contours, ice features, the quality of what's underfoot, and the wind. The Inuit can visualize large expanses of landscape as a map, but a significant part of their skill is related to language. Their vocabulary compels geometrical precision, and hence influences observation of land and ice. Thus the Inuit do not simply say that a rock projects from the snow "over there." They have to say "over there and up" (or "down," or "on the level").[26]

In Australia, the aboriginal people used song in a similarly precise way, with tune, rhythm, and words combining to describe the topography of vast deserts, conceived in terms of distinct traverses, each defined by its own song. But the songs are music and poetry as well, evoking memories of that particular landscape, and what the Ancestor did there.[27] Indeed, most traditional systems of geographical knowledge incorporate expressions of memory, values, and feelings, as Western Apache place-names in Arizona do, for example.[28] That is what distinguishes these knowledge systems as participatory.

So although it is well worth enquiring how traditional navigation functioned, we deceive ourselves if we think that Caroline Islander astronomy or Inuit snowscape specifications can be wholly translated into the language of scientific discourse. The fact is, these systems of navigation and geography are more than scientific knowledge, and carry other meanings to do with sense of place, and with life in a particular landscape. That is shown by what happens when people are displaced from their traditional way of life, or are forced to leave their traditional territory. Whereas they might adapt knowledge based on scientific study to new surroundings, people who depend on participatory experience rather than the knowledge achieved by detached minds can be fundamentally disoriented if transferred to a fresh environment. When people from hunter-gatherer communities have lost their land to colonists, or because of alien concepts of land tenure, they have commonly been engulfed by a terrible loss of meaning. Breakdowns and suicides become more common, people turn to drink, and there are community-wide dislocations. One person involved in this kind of situation has said: "We feel you are wanting to take away the spirit life . . . if you take away the power to control this land."[29]

In South America, where indigenous forest dwellers have been displaced by gold diggers, road builders and cattle ranchers, the result is to compel people to live "in a profound state of disharmony." Moving from forest villages to live in the poverty-stricken fringes of Lima, Bogota, La Paz, and São Paolo, "they have lost the meaning of their lives, the memory of the creation of the world." The Brazilian Indian who spoke these words makes the point that it is not only the injustice of losing their land that hurts, but the human and ecological disharmony.[30]

It would be easy to feel that although many such "backward" people have experienced great trauma, which illustrates the strength of their attachment to a place, personal and emotional upheavals have always been a part of modernization. But we should also note that for many such people, historical episodes with moral implications are remembered by the places where they occurred, and those who fail to remember the names of those places—hills, crags, trees—forget their own history also.[31]

Thus the sense of place may sometimes be linked to memories of local ecological disasters, and incorporate generations of experience that may have taught people how to live within limits set by nature. If modernization consisted of the careful use of science to show how to live fuller lives within those same limits, modernization could be very welcome. But when we observe a ruthless process of uprooting peoples, undermining their quality of life, and discarding their memory and experience, then we should be reminded of how readily we forget our own history, in North America and Europe, of dust bowls and other ecological catastrophes. The latter perhaps seem to include merely local disasters in which few people actually died. But there are also less frequent episodes of demographic collapse such as that in fourteenth-century Europe, when a long period of overexploitation and misuse of land was a contributory factor, prior to the epidemic of bubonic plague.[32]

Many people in the West lack awareness of this, partly because our own traditions of writing history have been dedicated to celebrating progress rather than recording warnings. But in addition, our inheritance of a mechanistic worldview gives little scope for us to acknowledge participatory experience, and warnings that might have come from that. Yet even as we deny the reality of such things, many of us, on another level, still tend to feel deep meanings in landscape and nature. One indication is the sense of mourning, loss, even depression that Hamilton-Paterson detects among people living in landscapes being despoiled by industrialization, house building, or road construction.[33] Another is that a few people, most conspicuously artists, seem not fully themselves in alien surroundings. Not only do they have a strong sense of place, but as with the indigenous peoples just quoted, personal identity for them seems to derive something from landscape.

In North America, for example, there are celebrated poets of place such as John Steinbeck in relation to the Salinas Valley, Faulkner in Mississippi, and Frost in New Hampshire and Vermont. And today, Wendell Berry is well known, among environmentalists at least, for the novels and poetry he writes about his corner of Kentucky.[34] There is also Harold Horwood, author of a stunning, celebratory book about life on the coast of Nova Scotia (Canada), where he finds a "sense of contentment, a sense of being in a place where one wants to be. . . . Here you could well believe that man and the world grew up together, perfectly suited and matched."[35]

By contrast, Margaret Atwood's sense of Canadian landscapes is of their inhospitable character. There are problems "in acceptance of the land" such that the deserted farmstead is an important symbol. Meanwhile, Dennis Lee explores the inflections of being Canadian in another way, stressing the importance of occupying "imaginatively and with integrity, one's own life and land," because if we live in a place that is radically in question for us, "that makes our barest speaking a problem."[36]

For me, English examples are the most vivid, and especially the experiences of visual artists. Thus John Constable produced many of his best paintings in the landscape he had known as a child and to which he constantly returned. Reflecting the immediacy of his visual experience, there was meaning for Constable in minor details: "Willows, Old rotten planks, slimy posts & brickwork, I love such things. . . . As long as I do paint I shall never cease to paint such places." Similarly, Robin Tanner, artist and etcher, would pick out details of a scene: "finely forged gate handles," or a "magnificent ashwood hay rake." But if these were not part of his home area in Wiltshire, and did not fit his sense of place, "something came between these things and me."[37]

Other people also seem to discover what their lives mean partly through attachment to a home territory. L. T. C. Rolt, onetime engineer, found this in the hills of the Welsh borders, and any similar hills aroused "strange exaltation" in him. Arthur Ransome recorded that whenever he returned to his home ground close to Coniston Water after a long absence, he would go to the shore of the lake and in a personal ritual, "dip my hand in the water."[38]

Farmers might seem to have greater reason to identify with the landscape where they live than any of these writers and artists, but when detached, economic attitudes to agriculture as a technology prevail, not all do so. On the Grey Prairie of Illinois, farmers of German descent tend to be concerned with continuity of landholding, regarding ownership of land as a sacred trust to be passed on within the family. That is a philosophy of place, encouraging a mixed farming strategy to maximize security, if not income. It involves a shared commitment of time from several members of a typical farming family. By contrast, Yankee farmers, of English descent, are more commercially oriented and entrepreneurial, regarding land as a commodity and agriculture as a wealth-creating business. The land on Yankee farms is predominantly given over to grain crops, and there is little livestock of any kind. There is greater concern to maximize financial returns, but less emphasis on "preserving soils for future generations."[39] Detached attitudes dominate.

However, it is not only farmers who have strong feelings about land. A comment on the urban scene in America notes how modern people look nostalgically to former rural lifestyles, yet are unwilling to sacrifice the comfort, convenience, and cash that they find in the cities. So they attempt to hang on to the spiritual value of nature in the modified "arcadia" of the leafy suburb. Around 1900, when streetcars linked suburbs to the city, it was said that technology was "putting arcadia within reach of city dwellers who would otherwise be denied its moral benefits."[40]

Many urban dwellers develop a sense of place around purely man-made landmarks, and the links between place and nature are then broken. Works of architecture or engineering rather than hills, trees, and lakes become the most prominent aspect of people's surroundings. Thus apart from designing suburbs in which trees and flowers are ever-present reminders of pastoral landscapes, there is also in our culture an enthusiasm for works of engineering and urban/industrial development that create new kinds of landscape or impressive spectacles within the existing scene.

Marking Land and Cherishing Nature

Historians of science often talk as if there is an unambiguous distinction between the detached, disenchanted worldview inherited from the scientific revolution of seventeenth century Europe, and the more "primitive," organic view of nature that preceded it. One cannot deny a major change in habits of thought that may be dated from about then. But throughout this chapter we have noticed that modern people have feelings about nature and place that seem to represent a lingering residue of an earlier outlook. Even the most disenchanted and scientifically minded modern person quite often comes to identify with a specific point on the landscape and feels that he or she has put down roots there.

In other ways, apart from the way we develop a sense of place, landscape seems to invite responses from us. I discuss three kinds of response in particular, of which the first two receive fuller attention in the next chapter.

First, an impulse to mark the landscape seems an integral part of the sense of place, as we noted on the first page of this chapter, with reference to the grave in the Matopo Hills, and Monkey's insistence on marking the pillars at the end of the universe. Whether or not it is appropriate to compare it with the way animals mark breeding territories with their scent, this impulse certainly has a long history in human cultures. That is especially well illustrated by the rock paintings and carvings to be found in many parts of Australia, Africa, and Europe, some very ancient. In all these regions, there have been peoples who modified places important to them by leaving marks that alter the earth. Their paintings or carvings are likely to have been connected with rituals or commemorations, sometimes connected with territorial claims or hunting or ancestors.

Although some of these marks—particularly the paintings—were the work of hunter-gatherer peoples, it is noticeable that these groups tended to mark the landscape only lightly. Richard Bradley argues that this is not because they lacked the capability or numbers to build larger monuments, but may reflect an attitude of respect for the land, or of feeling part of nature. The development of agriculture was associated with a changed attitude to landscape, Bradley argues. It was associated also with larger monuments for burial or ritual. In Europe, some of the most striking rock art was created when agriculture and pastoralism were still in their infancy, and hillside carvings seem to mark the furthest outposts of settlement or summer grazing.[41] Later, as much larger tracts of landscape were laid out with fields and houses, such ritual marking of the land no longer had much point. Buildings were more prominent landmarks, and later, in the medieval landscape, the church spire was a powerful pointer. Today, works of engineering as well as a great diversity of other structures mark the land, and we are conscious of the need for ritual marking only at the furthest limits of endeavor, as when we plant a flag on a mountain summit, or at one of the poles—or on the moon.

If these kinds of marking are one characteristically human response to landscape, a second response is the impulse to explore every detail of the place with which we identify as well as to adventure beyond its boundaries. Mumford comments that if "boundless oceans, starry skies, had not awakened his (or her) mind . . . the human would have been a very different creature."[42]

Third, though, many of us feel that we specially cherish certain features of our home ground. Sometimes also we encounter living things that seem so delicate and fragile that we feel drawn to protect them. That feeling may have stimulated the domestication of plants in ancient times. Today it is reflected in the houseplants, window boxes, or flower borders that many people maintain—and the animals they keep. Some of the peoples in South America and Australia quoted earlier cherish their lands as they cherish the communities to which they belong: The two things interlock. For a few artists also, the sense of place may be almost equally pronounced. Jane Howarth emphasizes the word "cherishing" in describing ways of valuing nature and place that carry these various connotations of caring for and protecting.[43]

Just as Part 1 of this book argued that technology by itself is better appreciated if we pay attention to human responses, so now we see that human impacts on the environment are better understood if we are aware of people's responses to nature and place. Equally, the work of conservationists is better informed if they understand the various ways in which people cherish plants, birds, and animals, even apart from their ecological significance.

One reason for valuing nature may be practical. People who are sick, or individuals who are stressed or suffering breakdowns, are found to benefit just by watching clouds drift across the sky, by seeing the slow changes in a growing plant or a bud bursting into flower. To enjoy such things can be to retune to a steadier pace of life. The seashore is especially good for retuning, because the expansiveness of the horizon, where it dissolves into sky, and the light glinting on water, combine with so many rhythmical experiences: waves, tides, and the flight of the many birds inhabiting coastal places.[44]

But we should be wary of valuing nature only as an aid to health, especially when we notice social scientists writing in manipulative language about measuring "quality of environment . . . by its capacity to promote behavioral or economic goals." These experts comment on how the importance of natural environment "in maintaining self-identity is firmly established in the psychological literature."[45] But one may still agree with Keith Basso that this analysis is too much rooted in a materialist, use-oriented attitude. Thinking especially of the Western Apache people, he comments that human groups everywhere "maintain a complex array of symbolic relationships with their physical surroundings . . . which may have little to do with the serious business of making a living." Scientists committed to measuring statistical regularities tend to miss this, because they regard the semiotic dimensions of the environment as epiphenomena, and they lack real interest "in what human beings take their environment to mean."[46]

Yet to understand what aspects of the environment are most strongly cherished and why, "should become part of our knowledge of human beings," Basso argues. Some ecologists know this, and have learned from the attitudes of indigenous peoples to their environments. However, what they have learned is usually formulated on a systemic level of abstraction, "well removed from the level of the individual." Basso then reminds us that "it is individuals, not social institutions," who make and act on the meanings of landscape and nature.[47]

Some thinkers, though, have learned more positively from the kinds of experience Basso documents. They talk much about "deep ecology," and seek a holistic philosophy that would integrate modern science into a more rounded approach to the understanding and cherishing of nature. Some who take this view suggest that "mind" or "self" is not a quality limited to humans and a few higher animals, but has ramifications for all the natural world.[48] Except in its most naive manifestations, this is not an attempt to reinvent nature as spirit, nor to reinvent God, but it could tend toward thinking of nature in terms of explicit purposes working themselves out.

That seems to me a dangerously overelaborate way of explaining why humans have a sense of being part of nature, with attachments to natural places. We do not need such elaborate explanations, because biologically, we are of nature, and as Chapter 1 argued, some of our sensibilities relate to processes and rhythms found throughout nature. The latter are related also to our sense of purposiveness and direction in life and are often reflected in music. But links between ourselves and nature are evident not only in our awareness of life's rhythms, but also, in a different way, in human responses to place.

In today's world, there is perhaps an increased sensitivity to nature among a minority who campaign to protect the environment, who study and enjoy the living things around them, and who celebrate their sense of place. In my own locality, there are "field days" in springtime during which people walk through and record their local landscape, giving expression thereby to their sense of attachment to it.

For the majority in modern consumer society, though, it is easy to feel that relationships that involve cherishing nature and place have all but disappeared. Many people prefer machines that express domination over nature through their noise and power: four-wheel-drive vehicles or speed-boats, for example. Others seem to have turned their back on nature altogether to live in an electronically mediated world. The digital revolution, Mark Slouka remarks, demands that we should "move indoors to renounce the external world," because technology is now seen as the "new nature," with virtual reality (VR) regarded as more exciting, more "real" even, than what is dismissively denoted as RL (for "real life").[49]

But people are also withdrawing indoors "because the world outside our homes has less and less to offer," due to the decline in quality of life noted earlier. Along a major highway in California, Slouka notes numerous communities with "identical (and very expensive) houses . . . each with a two-car garage. The postage-stamp lawns are manicured, perfect and empty . . . no life outside the home is possible here. There is no playground, no park, no field or meadow."[50]

This way of treating the environment is characterized as "de-creation" by Hamilton Paterson, who describes an island in the Philippines that Japanese companies have de-created to make it into a holiday resort served by helicopters, hydrofoils, and high-tension lines.[51] The process is being actively pursued all over the world, and like the other authors quoted, Hamilton-Paterson discusses it with immense feelings of loss. Contact with landscape and nature that once contributed meaning to people's lives is drastically reduced. When people are not visiting Disneyland or commodified holiday resorts, what is left for them to do but live indoors, with their home entertainment systems and virtual pets?

The new lifestyle provides many opportunities for making money on a grand scale, and much of that money translates into power over media empires, and over the shape of the electronic worlds now coming into being. It is in those worlds that we are now expected to locate our sense of place. But as the next chapter suggests, there are other options with regard to nature apart from turning our backs on it and then de-creating it.

6

Exploration, Invention, and the Remaking of Nature

Invitations from Nature

Thomas Jefferson's book *Notes on the State of Virginia,* begun in 1780, is mainly a factual account of the economy and government of his own home state. But in some passages, strong feelings emerge about the land as a source of meaning, and even of virtue. He held that America had a unique opportunity "with such a country before us to fill with people and with happiness," and with such "an immensity of land courting the industry of the husbandman."[1]

The word "courting" here is especially appropriate in expressing a part of human experience of landscape and nature, for we can feel so strongly drawn to specific places, or to specific activities within the landscape, that it is as if nature were indeed "courting" us, or "inviting"[2] our participation. Reflection on my own responses leads me to associate a landscape not seen before with feelings of anticipation, and definitely, with being invited to explore, or to linger and even settle. Readers have challenged the appropriateness of this language, but if I am to explain what I often feel about landscape, words about being invited or courted are those that come to mind. Equally, some places, such as the tops of mountains, can invite one to leave a mark: another stone added to the summit cairn, perhaps, or initials scratched upon a rock.

Feelings like this may relate to the sense of place discussed in Chapter 5 in either of two ways. We may feel invited to use, cultivate, or explore the nooks and crannies of a place we already know well, and to which we are already attached. Or the newness of an unfamiliar territory, or even the arrival of spring, may awaken an impulse to go further, exploring the unknown. Mabel Shaw, living in Central Africa in the 1930s, expressed this second feeling by commenting that in the first days of the dry season, the "sting and sparkle, freshness and fragrance" of early morning "filled one's inmost being with a strong wanderlust; to be on the road; to see Lake Tanganyika lying like a dream of still loveliness; to pitch one's tent in the vast forest."[3]

If this is seen as an authentic human response, it may need to be understood as arising from *participatory* experience of landscape, using this term as it was defined in Chapter 3. Then the contrast is quite clear with the *detached,* analytical style we often prefer.

Biologists and ethologists emphasize that an exploratory drive is part of everybody's makeup, and is present in animals also. It is an urge as basic as hunger, and is easily observed in laboratory animals and domestic pets. Exploratory behaviour is especially prominent when mammals are young, as they begin to learn what their surroundings offer in terms of food or shelter—and what hazards they need to avoid. On this level, exploration is part of the play behaviour of animals and humans that was discussed in Chapter 4 (where bibliographical references are found). There we saw that playful exploration can lead to collecting and classifying objects from the environment. But it can also include a ruthless curiosity, as when a child pulls some legs off a spider to see if the creature can still walk, or captures a butterfly and detaches its wings.

We have already noted that primeval humans did not easily live "in harmony with nature," nor do children. Rather, as they grow up, they find themselves increasingly moved by conflicting impulses. The sense of place and of identity with a home territory is in tension with an urge to explore way beyond that territory's limits. The impulse to protect and cherish small animals, flowers, gardens—perhaps whole ecosystems—is in tension with a destructive curiosity about nature. It may be in tension, too, with the need to use natural resources, and sometimes with aggressive urges to hunt or exploit. One expression of that tension is that some hunter-gatherers had rituals for asking forgiveness of the animals they killed.

Not only are we more aware of conflicting impulses as we grow older, but the way we resolve tensions among them may change. A retired British politician who is now prominent in movements to protect the countryside recalls how, as a boy, he enjoyed shooting starlings with an air gun, until one day he saw a bird he had injured writhing in agony, and was too upset to shoot any more.[4]

Even William Wordsworth, whose poetry so strongly suggests harmony with nature, admitted that in boyhood he took eggs from birds' nests, and set snares to catch woodcock. Going out late to see what he had caught, he felt "a trouble to the peace" of the starlit night. He also occasionally took a bird trapped by "another's toil," that is, in another man's snare, and that "Became my prey."[5]

Wordsworth writes of this as if it were one of "the coarser pleasures of my boyhood days." Other writers see it as a phase in the childhood of most boys (rather than girls).[6] One might guess that it was an impulse that, to a degree, persisted into adult life in former hunter-gatherer societies, but that, as with the two examples quoted here, it is an impulse that many modern people grow out of. However, for a significant minority, destructive impulses not only persist and influence attitudes to nature, but may be reflected in attitudes to people also (as we shall see in Chapter 8).

But interest in other animals was never limited to the destructive activities of boys who killed birds or insects. There has always been admiration as well for animals that could run very fast, swim well, or fly. When the horse was domesticated during the Bronze Age, its speed when running seems to have been the quality that people most envied and wished to appropriate for themselves. About 2000 B.C. in the Middle East, a pair of horses harnessed to a chariot could enable men to travel at speeds never before experienced. So the sun god, traversing the heavens each day from horizon to horizon, was imagined to be drawn by horses. And here, as in so many branches of technology, invention that appealed to the imagination (or was useful in warfare) preceded practical, utilitarian developments. Harness that enabled the horse to be used for heavy haulage, or to be saddled for easy riding, developed much later than the chariot.

The flight of birds had immense imaginative appeal in most cultures, and there were many legends about people who attempted to fly. It is wrong to assume that humans invented flying only in the twentieth century. "Man has always been airborne in his imagination."[7] In China, kites large enough to lift people were made centuries ago, and in the West, practical balloons were invented before 1800. People also experimented with wings, at first trying to make them flap. In the 1890s, though, Lilienthal showed how the principles of gliding could be used. In the next decade, another aviation pioneer developed his ideas about aircraft design in part by watching an albatross that glided with motionless wings above a ship he was on in the South Atlantic.[8]

The antiquity of the impulse to fly has sometimes been recognized in a limited, literary way. In the 1920s, an author who referred to the new power of "mechanical flight" commented on how often this was described by allusion to the old story of Icarus,[9] whose father made wings for himself and his son to fly from Crete to Greece. Arthur Koestler also commented on basic themes that keep cropping up in fiction and myth, and talks about ancient and persistent preoccupations that psychoanalysts have discussed in terms of "archetypes."[10] These are themes that connect with something "obscure and latent" going back beyond all modern expressions of technology, one example being the struggle to wrest power from the gods. This the legendary Prometheus did when he stole fire and gave it to man—and then was punished by being chained to a rock. Some historians have developed nice metaphors for the modern age of rapid technological progress—the period since the start of the industrial revolution—by asking: How did Prometheus escape from his chains? Who unbound him and released his creative energy? How did he enable humans to escape the inhibitions that had previously limited their inventiveness? The answer Prometheus himself gave to the last question was: "I sent blind hopes to settle (human) hearts."[11]

In discussing this archetypal struggle to control fire and all its power, Koestler mentioned many parallel legends, including the story of Adam eating of the tree of knowledge and more recent legends, such as that of Faust. He noted that these stories all describe human efforts to acquire power over nature, and they all offer warnings about the dangers of such an enterprise. Not only was Prometheus punished, but Icarus flew too near the sun, and waxen components in his wings melted.[12]

Although the search for Promethean power may become an obsession for some people (including the builders of bombs and rockets), obtaining more limited powers of motion or flight can be liberating in an innocent, enjoyable way. To set off on a journey and be able to choose one's speed—walking, cycling, riding a horse, driving a car—is to fulfill one's sense of individual capability and freedom. The very feeling of motion becomes a pleasure to be enjoyed.

Part of this enjoyment may again belong to our animal inheritance. When otters (for example) are playing, if they find a steep bank, wet and slippery after rain, they may slide down it, then run round and slide again repeatedly. Motion such as that is in itself enjoyable. Animal play may be explained as a process of refining muscular skills so that controlled but rapid motion is possible when needed for hunting, or to run from danger. Human sports and games can perform a similar function and are undoubtedly enjoyable too. The availability of horses, chariots, bicycles, and now cars can enable us to dramatize and reenact pleasures of motion and control first experienced in play.

Some of the ideas that are common currency regarding inventions such as the bicycle and automobile are influenced by rhetoric about the impact these inventions have had on society, and the way this has determined patterns of social change. In many instances, though, this form of words puts matters the wrong way around. Many inventions arise from the impulse to play, the enjoyment of motion, and the sense of being invited by nature to explore or imitate. It is these impulses that are the sources of the impacts discussed. It is they that are the causes of change, if we must speak in causal terms.

Similarly, in the modern world of computers, we can observe play and exploration in users' behaviour, and a sense of liberation. Here also, much is said about the impact of computerization, as if we were dealing with something that has come on us like a meteorite from nobody knows where. The reality is that the source of this technology is as much human as other major inventions. Like literacy, printing, firearms, bicycles, and automobiles, computers are self-revealing inventions. It is what we learn from them about ourselves—our impulses, purposes, abilities, and potential—that makes these technologies seem revolutionary.

Explorations and Journeys

Although human responses to nature may include impulses we can recognize also in playing otters and galloping horses, or in a human desire to fly like birds, one of the strongest impulses is that which makes us wish to explore the world and undertake hazardous journeys. Although this is an impulse that individuals in most human groups have experienced from time to time, people have varied greatly in how they explain it to themselves. In some societies people "went walkabout," and in others they went on pilgrimages. Christopher Columbus thought of his own explorations in mystical, often Biblical terms, sometimes seeing himself as a latter-day Noah.[13]

Captain James Cook, by contrast, was much more like the prosaic and rational investigator that a scientist is supposed to be. He was given to few expressions of feeling, and had a specific, scientific objective for his first voyage: to observe the 1769 transit of Venus from Tahiti. He was like a scientist also in that "nothing . . . gave him greater satisfaction than exploding myths and establishing truth," notably about the Great Continent that some had supposed must exist in the South Pacific. In that respect, Cook's greatest achievement was to prove a negative.[14]

Underneath his reserve, though, Cook was driven by restless energy and a willingness to persist with possibilities that others had not the courage or vigor to pursue. In January 1774, when his ships were at their furthest point south in Antarctic seas, Cook was "not sorry" that ice blocked the way into even more inhospitable regions. Significantly, too, he admitted that "ambition" had led him so far, and that this was "not only farther than any other man has been before me, but as far as I think it possible for man to go."[15]

Historians seem at a loss, however, to explain the ambition of explorers, especially those nineteenth-century men (and some women) whose expeditions into the unknown (as Europeans saw it) seem to defy all reason. In the exploration of Africa, for example, there is little clarity in any account about the motivations of individuals, some of which, indeed, seem to reflect "purposiveness without purpose." But Alan Moorhead offers two significant comments. First, many of the explorers seem to have been "born with something lacking in their lives," and experienced "a fundamental restlessness." Second, some felt "impelled to go back again and again." Yet they were rarely touched by the beauty or grandeur of the African landscape. It was all seen as "hostile, incomplete, not to be regarded with an aesthetic eye until . . . reformed and reduced to order."[16]

By contrast, English explorers may have been attracted to the polar regions by the awesomeness yet tranquility that icebound landscapes inspired. However, English expeditions were often characterized by "poignant absurdity" and incompetence. Whereas Scandinavians such as Amundsen were glad to learn from the Inuit inhabitants of the Arctic how to travel, hunt, and fish in that terrain, English explorers sometimes starved as a result of their contempt for Inuit methods.[17]

That was especially and tragically true of Franklin's search for a northwest passage through arctic seas north of the Americas, in which his ships were crushed by ice and men died of hunger. Was there not some purpose that apologists for this venture should have acknowledged, apart from the commercial value of a northwestern route to Asia, if one should be found?

Having posed the question, an otherwise unremarkable book on arctic exploration points to motives relevant not just to exploration but to other aspects of science and technology, speaking of "the poetry, almost the mysticism, behind the long search." Once a problem is set, its solution becomes an imperative, "as Everest soars and must be climbed." Behind the scientific curiosity in exploration lurks something "harder and more primitive, something that can make myths, found systems of thought, and people the empty seas." Herbert Read is quoted as speaking of moments when an artist "is carried beyond his rational self, onto another ethical plane." The quest for the Northwest Passage was "so extraordinary a phenomenon of the human spirit" that it must be seen in those terms.[18]

The deficiency of this account is that it sees only nobility in what might otherwise be regarded as a destructive obsession, and does not recognize the negative aspect of quests and imperatives. Another author, writing about Ranulph Fiennes, a modern adventurer who has walked to both the South and North Poles, wondered if he is driven by a wish to be always testing himself. Linked to that, "something fundamental is missing—a lack of interest in and understanding of other human beings."[19] Similar things were said about Jean Batten, a pilot who made record-breaking solo flights between Australia and England in 1934 and 1936. She seems to have been entirely absorbed by her enthusiasm for flying, and was "the greatest navigator and all-round aviator of her day." Yet her life was a "lonely tragedy." It seems almost that her achievements were an outcome of that loneliness.[20]

Noting how people of comparable personal character become involved in maritime exploration of an "obsessive" kind, Hamilton-Paterson wrote of Robert Ballard's search for the wreck of the *Titanic* as a quest pursued with such determination that it was as if some "private thing" had been lost, not just a shipwreck—as if he were searching some "psychic deep" within himself.[21] A clue to what might be missing, and what is being searched for, is again that many of these adventurers seem to have lacked understanding of the more intimate side of life.

There may be a connection here with the findings of psychologists quoted in Chapter 2 that some individuals drawn to work in engineering appear to be slightly autistic, and prefer research with an object-centered focus. It could be that some people became explorers in the nineteenth century for similar reasons. They had a greater interest in the physical shape of continents than in the people inhabiting them, and maybe were drawn to polar regions because there was nobody else there. Solo flights and voyages would have a similar appeal. Among explorers, as among scientists and inventors, a compulsive interest in a project or "quest" does therefore seem to be one direction in which object-centered interests can take a person.

Remaking the Landscape

When Thomas Jefferson wrote of America as a land "courting the industry of the husbandman," he was thinking, quite clearly, of wild landscapes being tamed and used for agriculture. He did not envisage such a drastic remaking of the landscape as we so often encounter today, when whole tracts of countryside can disappear under the concrete of freeways and flyovers, dams or urban sprawl. In many people's experience, technology has largely displaced nature in the immediate environment of their lives. Ezra Pound expressed the positive side of this displacement when he saw New York lit up at night: "Here is our poetry, for we have pullled down the stars to our will."[22] But half a century after Pound, Jacques Ellul put the matter in a different perspective by remarking that the current aim of civilization was to replace the "natural milieu" of people's lives with a "technical milieu" in which "everything that goes . . . to make livelihood, habitat and habit is modified."[23]

The technical milieu has become a reality since Ellul wrote to a quite extraordinary extent, partly through the alteration of landscape, but partly also as the electronic media have become so prominent that they seem to become an alternative world to which some people withdraw. That raises again issues that emerged toward the end of the previous chapter. The human impulse to mark the landscape was originally a response to the sense of place and a primitive need to demarcate territory. But land use is now so intensive that in many places, it has begun to extinguish the human meanings associated with place. Questions need to be asked about different ways of using land and the balance among them. But for some people, the point of balance has long been passed, and the conditions of their lives are depressing to the human spirit.

To present the modern environmental crisis in terms of low morale and loss of meaning is not the usual approach, though. More commonly, the crisis is seen as a question of biodiversity, pollution, or resources. The focus of this book on matters of personal experience and existential meaning may seem much less important. Yet the economic and ecological degradation of the environment has a counterpart in human experience of alienation and loss that needs to be recognized. Indeed, the remaking of the world as a technical milieu—and now the remaking of the genetic basis of life—raises urgent questions on every level: existential, social, and economic, as well as ecological.

As some people see it, the drive to replace nature, at least partly, with a technical milieu is the great modern gamble. The question they have in mind is whether this new order is something we can support over a long period. Is it sustainable? Can we maintain the production of crops, energy and other essentials in a world where many natural processes have been modified or replaced? This is the bet of the century—the twenty-first century—not only because of the risks inherent in replacing natural systems, but also because the aim is not a new equilibrium, but a world of continuous change, equated with technical progress and economic growth. Associated with this is the attitude that if there are problems with our technical world, we need more technology, not less, to solve them.[24]

The latter point has to be considered in the knowledge that air pollution is already altering weather systems throughout the world, and that the extinction of plant and animal species is becoming as momentous as the great extinction associated with the disappearance of the dinosaurs. Few wilderness areas remain unaffected by human activities, with tourists and refuse tips now even in Antarctica (although fortunately, there is a fifty-year ban, from 1998, on mining and oil exploration south of latitude 60°S).

Even so, Bill McKibben is largely correct in saying that "the separate and wild province, the world apart from man," has been gravely compromised. Or as others have said, we have created a world in which people find themselves "bound fast in a new ice age of technology and bureaucracy" in which shallow optimism and synthetic scenery are provided by Disneyism in all its manifestations, but real nature is hard to find.[25]

Another kind of synthetic world, though more transient and also more thought-provoking with regard to how people feel about transformations of landscape, is suggested by the artist Christo, who has explored the significance of human marks on the land, and on monuments in towns, with his famous plastic curtains and wraps. More soberly, Richard Long and Andy Goldsworthy are artists who have investigated the meaning of landscape by making their own patterns with stones, twigs, branches, or leaves on smooth beaches or grassy hillsides. These tend to demonstrate human modifications of the landscape that "feel" appropriate,[26] just as some painters and poets portray landscapes with human-made fields and roads that seem fitting and even beautiful.

In this context, the civil engineer can rightly feel that his or her constructions have potential to add meaning to the terrain, rather than, as critics may say, despoil it. Indeed, the engineer can point to a tradition of feeling that it is proper and right for humans to leave their mark on the land; that landscape can be charged with meaning,[27] and that nature can be "hallowed" by human activity. As one modern poet says:

Nothing but human use can glorify
field, mist, air or light
common possession and the common right.[28]

But there is considerable tension between different views about this. "To some people, a river valley is incomplete, unfulfilled" if it is not traversed by a road or flooded by a dam. "To others the opposite holds." Part of the problem is that in many places, there is already too much development. One line of electricity pylons can be thrilling, like a row of giants stalking the land. But a network of pylons and cables makes the countryside a prison camp, trapping us in the concrete jungle that so often spreads rampantly around the pylons' feet. We are faced with the vanishing of entire landscapes, and it is this that "threatens us most" as on one Pacific island that has lost all its indigenous birds, and "the quietness of death reigns where all was melody."[29] And the destruction of forest landscapes in South America and Africa means that fewer migrant birds return north each year. In Europe, as in North America, the noise of road traffic more than ever replaces birdsong as the predominant rural sound.

Yet it has been widely accepted as permissible and appropriate for large parts of the natural landscape to be entirely replaced by a man-made technological environment. The development of cities presupposes this for limited areas, but industrial societies take over many other areas for transport infrastructure, mines, and factories. The nineteenth-century industrial landscape, "with its cavernous factories draped in smoke" was quite often seen at the time as a legitimate expression of "man's new powers of transformation." It was understood in terms of the "technological sublime" as something that could rival or perhaps replace the sublime in nature.[30] Today, the smoke of that kind of environment is regarded with distaste, but not the principle of a wholly transformed landscape. Wilderness, forest, and farmland are giving way to cityscape and concrete jungle on every continent. It is necessary, then, for us to ask where the balance lies between ways of using land that are humanly and ecologically valid, and ways of marking and using it that both depress the human spirit and irreversibly destroy ecosystems.

Engineering and Gardening

If we look at different ways in which people have tried to define where the balance between nature and technology should lie, there is a range of ideas to consider, extending from the ancient art of geomancy through the ideal of the garden and the Enlightenment concept of a middle landscape to modern concepts of sustainability. Before considering these, however, it is worth noting two engineering approaches, Chinese and European.

One of the most eloquent expressions of the latter is to be found in the autobiography of L. T. C. Rolt, a British engineer who found great satisfactions in a career in mechanical engineering—building engines and harnessing the elemental forces of fire and steam—but who then felt appalled by the dirty, denatured industrial city that this activity had created, and by the impoverished lives of many of those employed there. Later, though, he found a happier balance between engineering and nature in the English canal system, whose waterways were small enough in scale to enhance rather than dominate the landscape, and whose earthworks and aquatic features provided many new niches in which wildlife could flourish. Some of the same things have been said about old canal systems elsewhere in the world, such as those of Lombok and Bali in Southeast Asia. This kind of engineering did not attempt to dominate or replace the natural world by an industrial one. It could express "harmony with nature."[31]

In another of his books, Rolt seems to identify himself with nineteenth-century engineer James Nasmyth when he was confronted with a bleak vision of industry during a visit to the English "Black Country." There, "the earth seemed to have been torn inside out. . . . Its entrails are strewn about . . . and the smoke of the ironworks hangs over it . . . Amidst these flaming, smoky, clanging works, I beheld the remains of what had once been happy farmhouses, now ruined and deserted . . . surrounded by clumps of trees, black and lifeless."[32]

Both Rolt and Nasmyth, through conflict within their own lives, exhibited the desire for technology to be used in ways that harmonize with rather than threaten nature. Both were gifted and enthusiastic engineers, yet were appalled by some of what engineering led to, and both retired from the engineering profession relatively young.

In China, over many centuries, a comparable dilemma about what harmonizes with nature and what does not was reflected in discussions between two schools of thought in hydraulic engineering: one favored "confining and repressing Nature," the other preferred "letting Nature take her course."

In his volume on civil engineering in China, Joseph Needham showed that engineers who took the latter view were mainly Daoist (Taoist) in philosophy, and where irrigation works were concerned, believed that the building of dams should be avoided, and that other structures should work in partnership with nature, such that "a good canal is scoured by its own water; a good embankment is consolidated by the sediment brought against it." The opposite view prevails now, in modern China, otherwise the high-risk Three Gorges dam would never have been contemplated.

Needham further described a great irrigation scheme in Sichuan province, built about 200 B.C., and capable of watering thousands of acres without resort to a "big dam" approach. Some long time after it was completed, two temples were built overlooking the headwaters of the main canal, to commemorate the engineer-administrators responsible for its construction. As Needham said: "The Chinese were never content to regard notable works of great benefit to the people from a purely utilitarian point of view." With their characteristic sensitivity to the significance of human marks on the landscape, and their ability "to raise the secular to the level of the numinous," they could see beyond practical engineering to deeper meanings. Moreover, the statues and inscriptions in the temples are not only of religious significance, and not only praise the builders, but they also include texts poetically setting out the engineering principles of deep channels and low spillways that the works embody.[33]

Such Daoist sentiment, which is not against technology, but which avoids the attempt to conquer nature by means of massive forms of construction, may be a philosophy that can be adapted to address some of our present dilemmas. The irrigation scheme that it celebrates, if accurately reported, is also an example of sustainability, having been in operation for more than 2,000 years.

Better known today is another Chinese tradition regarding land, complementary to the way of thinking just quoted. Sometimes referred to as geomancy, but also well-known by its Chinese name *feng shui* (which means "wind and water"), this can be compared with European traditions in alchemy (Chapter 3) to the extent that it refers to authentic participatory experience. As in alchemy, there is also a tendency to mystification, although the subject matter is land forms and "energies," rather than metals and "virtues." Undoubtedly, in many parts of China, *feng shui* helped create landscapes in which buildings were sited in a balanced visual relationship with hills and water (sometimes including artificial lakes, as at the Summer Palace west of Beijing).

One further way in which peoples of many cultures have expressed their feelings about the relation between human artifice and the natural world is by making gardens. It should come as no surprise, then, to notice that the Chinese have long been enthusiastic about gardens (in which water was often a feature), and about the study of botany and horticulture.[34]

Mumford stressed the importance of the garden during an early phase of human innovation, when plants were being domesticated and pottery was first made. Traditions established in this phase of human history may well linger in all cultures where horticulture and agriculture are practiced. One visitor to the famous and lovely garden in France that belonged to Monet, the Impressionist painter, saw it as an expression of widely shared values. "People of all nationalities, from all over the world, were wandering round, all understanding what they saw without need of interpretation. The love of human creativity and natural life in that garden was . . . palpable and overwhelming in its intensity."[35]

But the garden as a vision of gentle creativity and harmony with nature is not the only possibility. Much conventional gardening today aims at excessive tidiness and neatness through drastic overuse of chemicals. Historically, where the ideal of technology as controlling and overpowering nature was as influential as it is today, gardens were often strictly geometrical in layout and heavily dependent on mechanical technology. It is no coincidence that the great gardens of Europe during the period of the scientific revolution were of this kind, with the skills of hydraulic engineers reflected in their elaborate fountains (as at Versailles).

Medieval Islamic culture showed a similar mechanical emphasis. Gardens were places to escape from the scorching deserts of Syria, Arabia, and Iraq, and depended on a good deal of technical artifice to overcome this arid aspect of nature. Many references in Islamic poetry, and in the popular *Arabian Nights,* mention gardens "watered by crystal brooks," or "shaded by palm trees and refreshed by a gentle flowing stream" in which "apples, plums and quinces hang in clusters from the boughs." Always there was water and shade to make a welcome contrast with the harshness of the surrounding deserts, and much water was also needed to ensure the survival of fruit trees. Elaborate supply systems were designed using canals, aqueducts, and tunnels. Fountains were often contrived as garden monuments, and these frequently depended on lifting water to a high cistern using a wheel with a chain of pots or other mechanisms. In medieval Baghdad, the machine and the garden worked in partnership, and both were subjects of intellectual interest. Water engineering, with its aqueducts, header tanks, pipes, water-raising wheels, and occasional pumps, made the garden possible. A book written in A.D. 1206 mentions pumps with metal cylinders associated with designs for garden fountains. It is remarkable, indeed, how often the most demanding technical problems that engineers have had to solve relate to monuments rather than objects of utilitarian concern.[36]

But although Islamic gardens might require the use of elaborate technology, much of it would be hidden, and in the garden itself all one would see might be a fountain or pool. More expressive of the ideal of partnership between nature and technology is the garden into which some aspect of everyday technology is openly introduced. Today, many people do this without aesthetic intent by allowing a parked car to dominate their limited garden space. Others ornament their plots with items expressive of a lost rural lifestyle, such as old wagon wheels, barrows, or horse plows. Such gardens seem to be saying that there was once a form of technology that could be seen as a partnership with nature, but no longer.

More positive was the image of machines in a garden illustrated in a schoolbook of 1910, with models designed for teaching children about the principles used by different power sources: steam, wind, and water. Taken individually, many of the small machines represented could be seen as examples of human mastery over nature, but presented in a garden setting surrounded by big trees, they took their place beside flowers and a neatly mown lawn as portraying a balance between nature and artifice. For one reader of that schoolbook, at least, this garden implanted "a longing to participate in a world in which the works of nature and human kind do not conflict but complement each other."[37]

The Middle Landscape

Daoist engineers in China and gardeners in many cultures expressed a view about how technology should be used that had parallels with the ideal of a "middle landscape" discussed in the United States from about 1780. For example, one of Thomas Jefferson's correspondents characterized the western frontier of settlement as a place where men behaved "no better than carnivorous animals," but at the same time, he described Europe as possessing an oppressive society of great estates, and landless people in poverty. Midway between lay the good farmland and "fair cities" of the eastern parts of America, one region of which was described by another writer as a "middle state, between the *savage* and the *refined*." Here was a land of "substantial villages, extensive fields . . . decent houses, good roads, orchards, meadows, bridges." America was "a place apart—a peaceful, lovely, classless, bountiful pasture."[38]

This, then, was a "middle landscape" in which nature was modified, but not obliterated, by the creation of meadows and orchards. And it presented an ideal with which Jefferson greatly sympathized, even while he recognized that the industrial revolution was taking root in America. His book about Virginia expresses views on this that, we should note, incorporated a social ideal. He wanted to fill the country "with people and with happiness," and looked on farming as a morally improving way of life that would contribute to that goal.[39]

Jefferson admitted these views to be "theory only," but a pastoral ideal remained strongly alive in America. Leo Marx has argued that the idea of the continent's landscape as a garden—a scene of productive and virtuous labor—has stirred deeper feelings in American culture than the apparently more exciting frontier ideal of the Wild West. The middle landscape was the garden ideal in another guise. It was a province where "sufficiency" was emphasized more than economic growth, and where the husbandman was "free of the tyranny of the market."[40]

Gardens and farms of this kind express feelings of attachment to land, and they mark the landscape in a way that expresses the sense of place that an attachment brings. In those respects, there seems to be some continuity with primitive attitudes to landscape and nature. In other ways, however, the idea of harmony with nature expressed by a garden or middle landscape is quite different from the relationship with nature felt by many of the hunter-gatherer peoples mentioned earlier. For some of them, Nature was a world of spiritual activity—of the Earth mother, the Great Spirit, and the living spirits of animals and trees. By contrast, the people of the Enlightenment who spoke about middle landscape would regard nature as a world of impersonal forces. Farms and gardens were technologically contrived by countering those forces with ax and plow.

Moreover, the contrived garden was valued more than nature's garden. When European colonists first arrived in Virginia and other of the milder parts of North America, they encountered such a profusion of fruit, flowers, trees, and game that they sometimes felt they were already in a garden where "scarlet blankets of strawberries painted the bellies of (their) horses . . . and grapes bowered the streams and rivers." Frederick Turner commented that in describing it thus, if this was a garden, "the whites wanted it not as it was but only as they might remake it," by cutting back the trees, shooting the wildlife, and banishing the native peoples whose "nature religion" the Europeans found disturbing.[41]

The middle landscape was essentially a remade garden, harmonizing with nature to a degree, but artificial in its control of planting and wildlife and its use of machines. Moreover, the early phases of industrialization could often fit neatly into this middle landscape. The first factories were powered by waterwheels and had to be dispersed along the rivers. Usually they were not very large, but David Nye comments that "even the Amoskeag and Lowell factories, which reached impressive proportions, were at first perceived to be in harmony with the natural order." The steam-driven factories that came later more often "dominated their surroundings and were understood to be dynamic, unnatural environments."[42]

The middle landscape was in many respects the creation of "scientific consciousness," reflecting confidence in human control of nature, and human ability to improve on natural landscape. But yet there is a residue of feeling in the writings quoted, which implies the lingering influence of more traditional, participatory responses.

The same mixture may be encountered in the very different social context of a nineteenth-century Russian estate as it was described by

Tolstoy. On a day when the landowner was inspecting a new threshing machine, indeed, his thoughts were switching from participatory to scientific modes. He looked at the sunlight on the threshing floor, and "at the white-breasted swallows that flew chirping in under the roof . . . then at the peasants bustling in the dark dusty barn." What was the purpose of all this? Was it really just about producing grain to fill one's belly? The swallow seemed to indicate an answer—but then his mind reverted to its habitual, scientific way of thinking, and "he looked at his watch to reckon how much (was) threshed in an hour."[43]

Much discussion of agriculture in the West proceeds on the assumption that farming has only economic meaning: the kind of meaning with which Tolstoy's landowner was dealing when he timed the work of the threshers. Farmers are regarded as entrepreneurs whose land is merely an investment, and who plan their strategies for growing crops or raising livestock solely with a view to the best possible financial return. This ignores the way that farmers may be motivated by the social and personal meanings they find in their work. Far from trying to maximize financial returns, they may be thinking of the security of their families while at times making decisions on the basis of what they like doing, and what gives them satisfaction. As Tolstoy's landowner watched some peasants bringing a hay cart home, a woman "broke into song," and others joined her, their voices in unison. There ought to be room for satisfactions of this kind, Tolstoy implied.

If this were just a comfortably placed writer with a romantic view of agriculture, his point might not be worth our attention. But harvest celebrations were once widespread, and are mentioned also by those who write from a laborer's point of view (Chapter 5). At harvest time in Ireland, every wagonload of oats brought back into the stackyard was "like the end of an act of creation." After the last load, "elated and set free we began at once to make ready the Harvest Dance."[44] Tolstoy is surely right to show how satisfactions of this kind gave meaning to farming even while economic calculations were important and necessary. Similarly, Jefferson's interest in science and its application to farming coexisted with a strong sense of the social and moral meaning of agriculture. The middle landscape was not only (or even mainly) a way of thinking about farming in relation to nature. It also implied an ideal for society.

Currently, questions are often asked about agriculture of the kind that depends on chemical fertilizers and pesticides, elaborate machinery, and monocropping. Comparisons are made with various forms of agriculture that are said to be "sustainable," involving fewer (or no) chemicals and emphasizing mixed cropping (or indeed, mixed farming with livestock complementing crops). This can easily lead to a wholly technical discussion about what practices are sustainable in the long term, but there is sometimes another dimension to the debate as well. Those who feel concerned about the environmental implications of modern agriculture also tend to be uneasy about farmers who have no sense of place and appear alienated from local communities. There may seem to be a correlation between these rather detached attitudes and interest in the most modern techniques. By contrast, advocates of sustainable agriculture may start with ecological concerns that were hardly recognized before the twentieth century, but often come back to a quasi-Jeffersonian solution at the social level. Ideas about committed farmers, family holdings, and a gardenlike middle landscape tend to reappear.

For example, one book that gave technical detail about soil conservation, biodiversity and sustainable levels of energy use also presented agriculture as a "cultural activity that provides meaning, cultivates moral responsibility, and continues traditions of caring for the earth and future generations." The book showed why it is important to understand how human society, land management practices, and farm technologies can evolve together as a system that "values humans as well as the ecological components," and takes account of "environmental soundness, economic viability and social justice among all sectors of society."[45]

When it comes to the specifics of all this, the similarity with Jeffersonian ideals becomes very evident. Wendell Berry wrote "a defence of the family farm," and others have cited Amish, Mennonite, or German-descended farmers in the American Midwest and Canada as people who practice agriculture on a family basis, using techniques that approach sustainability (if not wholly, at least to a significant degree).[46] Such farmers, it becomes clear, create a diversified middle landscape even where they have not much considered the scientific or philosophical reasons for doing so. Moreover, they sometimes influence neighboring communities, occasionally negatively when they seem stuck in the past, but often positively through their example of self-help and environmental concern.

Modern Environments

During one of his trips into the virgin forests of Maine, Henry David Thoreau climbed Mount Katahdin, which at 5,268 feet (1,610 m) is the highest mountain in the state. Afterward he wrote: "Here was no man's garden. . . . It was not lawn, nor pasture, nor mead, nor woodland, nor lea, nor arable, nor waste-land. It was the fresh and natural surface of the planet Earth, as it was made forever and ever." That defines wilderness relative to middle landscape, as does Thoreau's comment that the vast forests of Maine were "inhuman," however beautiful, and "it was a relief to get back to our smooth but still varied landscape (in Massachusetts)."[47]

It sometimes seems that for many people in modern consumer societies, even middle landscape is too stark, the weather too variable, the necessity occasionally to walk too tiring. So they are happier relaxing indoors with their electronic entertainments. To take that attitude, though, is to say that land and nature no longer have meaning except as means to produce food and raw materials. We might as well leave living things to be engineered in whatever way scientists think will best enable the land to produce food for a growing population and profits for agroindustry. We might also just as well subscribe to the view that market forces will stimulate whatever innovations are required to keep us fed and clothed. If resources of some essential material or fuel begin to run short, the argument goes, prices will increase, and that will prompt inventive people and progressive companies to seek other materials to do the job, or find other sources of energy. Economists who think this way seem so impressed by human creativity that they believe people to have limitless capacity to invent new resources.

However, many aspects of the environment, including the atmosphere, soil structures, and biodiversity, are outside the scope of economics. So "a free-market approach to the global pollution crisis seems inherently impossible. No one owns the air or the water." With no private property for sale, there is no free-market price for clean air, and so no incentive to take measures that will keep it clean. Clever ways have been devised for getting around this, such as requiring every industry, household, or vehicle that causes pollution to have a permit before it may operate. If permits had to be bought, and were freely traded, their rising price could create market pressures that would tend to limit pollution.[48]

Something might well be gained this way, although most feasible schemes deal with only a fraction of the overall environmental problem—with pollution but not biodiversity, or with energy but not entropy. Modifications to industrial processes informed by the so-called natural step approach may take more account of these issues, but rarely the whole range.[49] Even then, answers on a technical level may be unrelated to the existential experience of people who feel alienated from nature, which may be a more serious part of the problem for all of us than we usually allow. For some communities, alienation from nature leads to abuse of the environment. For others, it is clearly a major source of unhappiness and ill health. Jerry Mander sees the fate of aboriginal peoples, such as those of South America and Australia discussed earlier, as a critical symptom. He also observes that Westerners lack the "sense of the sacred" possessed by many such people, and that as a result, our technology is too much oriented to "overpowering nature."[50]

It is striking, indeed, how many authors come back to ideas about the sense of the sacred or a reverential attitude to nature once the seriousness of the environmental crisis is recognized. For then it is apparent that this is not a crisis that can be dealt with merely by creating economic incentives to reduce pollution, nor by cleaning up industrial processes and using "environment-friendly" consumer products, however helpful such measures may be as a start. Changes in lifestyle and a fundamental redirection of values and goals are required also. Such changes, it seems to be thought, depend on recovery of the reverential.

For example, in discussing the alarming rate of extinction of animal species, Colin Tudge mentions human populations that eventually arrived at some degree of balance with the landscapes they inhabited. A sense developed among them that they shared those landscapes with the spirits of trees and animals. This made people sensitive about what they took from nature for food, fuel, clothing, and shelter. In other words, religious feeling seems to have been supportive of environmental values. But today, Tudge remarks, "we have largely abandoned religion," and some would add that the Judaeo-Christian tradition was anyway more likely to encourage exploitation rather than conservation. So there is a need, Tudge suggests, if not for a new religion, at least for attitudes that can perform its former function.[51]

Perhaps these new attitudes will derive from the philosophy of "deep ecology," as it has been expounded, for example, by Freya Mathews. She also wrote of the need for a reverent conservationist attitude and asked whether nature embodies "a spiritual principle." She then added that rituals of place stemming from the sense of attachment to landscape can contribute to ecological insight by making one aware of local detail, and the particulars of specific environments.[52]

Alan Drengson, another exponent of deep ecology, has argued in a comparable way that "humans are . . . meaning-creating beings" who need to invent myths and stories that convey values and meaning. Such myths are "vital for individuals and cultures." He then asks whether the "recovery of our larger visionary self" as it might be achieved through such mythmaking can be related to "technology practices so that they will be ecologically wise?"[53]

My own approach is somewhat different and more distrustful of modern myths and new religions. The field days held in my own locality might count as "rituals of place," but they comprise only walking, looking, and recording the landscape in which I live. Apart from that, one should not jump from recognizing the limitations of disenchanted materialism into the comfortable embrace of some reinvented religion.

Instead, I look for something more basic, namely an "affirmative way"[54] of keeping in touch with my own feelings, and of enjoying the wonderful vitality and musicality of nature, through visual and tactile experience and my sense of place, not least as the latter is expressed by gardens. Indeed, the garden, properly understood, could be a paradigm— a model—for all our dealings with nature, especially if we regard national parks, wildlife reserves and any field where nature is cherished as garden.

Michael Pollan discussed the garden as a place "where nature and culture can be wedded," and suggested an "ethic of garden" that would paradoxically "cultivate" wilderness while recognizing that humans need to modify landscape and attack pests and diseases to survive. The ambition of conquering the earth should be abandoned, he added, in favor of a more collaborative approach in which we borrow nature's methods, as in organic farming, and protect nature's diversity.[55]

With regard to diversity especially, even small suburban gardens can be surprisingly effective as refuges for wild animals and birds. In Britain, ornithologists with suburban gardens now record a wider range of species than their rural counterparts because of the damaging effects of chemicalized agriculture in many rural locations. White-tailed deer have flourished in the backyards of Cincinnati. A naturalistic garden near Nuremberg, Germany, has attracted 700 animal species (insects, birds, mammals), and a comparable garden in Leicester, England, has 1,800 (including some very rare insects).[56]

A more abstract way of looking at the issue, and of summarizing the argument, would be to see the garden as a place where the defined purposes of the human gardener, conservationist or farmer encounter the undefined purposiveness of nature. We have a choice between *either* imposing our own purposes without any compromise, *or* of understanding and working along with nature's own purposiveness.

There is a close analogy here with the way we encounter the purposiveness of nature in the rhythms of our own bodies, yet also have conscious goals for our lives. We can choose to force the pace and live a goal-driven life. That can lead to more stress than is good for our health, which may be compared with the effects of agricultural practices that seek to make nature conform to our patterns. Or we can periodically retune our lives to more natural rhythms, as suggested earlier in this chapter, by taking time to enjoy growing plants or to walk by the sea. Or, more fruitfully, we can find ways of combining our own defined purposes with a natural rhythm of life, as J. S. Bach did in music when he paced compositions to incorporate heartbeat and breathing rhythms while at the same time exploring mathematical patterns, emotional resonances, theological symbols—and anagrams on his own name.[57]

A garden, in the wide sense indicated earlier, can be the ecological analogue of that kind of music, allowing us to do most of what we want to do in agriculture and other technologies, but at an altogether different pace. A garden can be a paradigm for environmentally appropriate technology to set against the currently dominant paradigm that aims to remake nature and compel us to live entirely in a technical milieu.

7

Gender and Creativity

Human Meanings and Their Loss

When it comes to asking ultimate questions about the meaning of life, "Wise men write many books in words too hard to understand. But this, the purpose of our lives, the end of all our struggles, is beyond human wisdom."[1]

The troubled person who is asking that question sometimes thinks of home, and finds consolation in memories of mist on the surrounding mountains and the "deep melodious names" of nearby hills and rivers. But sometimes despair goes too deep to be relieved by such thoughts, and the hills are remembered only as "desolate beneath the pitiless sun." At such times, there is more comfort in thinking about friends, and in playing with an infant grandchild. Holding this baby is the greatest comfort of all, especially when the small, serious face relaxes into smiles. Then more questions come to mind: "Who indeed knows the secret of the earthly pilgrimage? Who knows for what we live and struggle and die? Who knows why the warm flesh of a child is such a comfort?"

There is a progression in this questioning from frustration with reasoning ("wise men" and "books"), to intuitive meanings found in landscape, and then to the more fulfilling meanings of human relationships. There is a progression, in other words, from sense of place to sense of person, and the fullest meaning is found in the open, unconditional smiles and bodily warmth of the baby. Like Thomas Traherne in the seventeenth century, the questioner seems to be asking: "Is it not strange that an infant should . . . see those mysteries that the books of the learned never unfold?"

A further connection of meanings in place to meanings in relationships was expressed by Traherne when he found sunlight illuminating the "beauty of hills and valleys" and "sprinkling flowers upon the ground," because of his love for another person. That made him exclaim, "We are made to love . . . and to answer the beauties in every creature."[2] Meanings in nature and in human relationships are linked, and for him seemed all-embracing. But novelist Elizabeth Goudge reproached herself for loving "places too much and people not enough," or more negatively, D. H. Lawrence used a landscape despoiled by coal workings as a metaphor for a world of relationships lacking in tenderness.[3]

One might think that if such meanings are really all-embracing, love for others must be able to inform work in technology. Some examples of this will be discussed in the final chapter, but they are found mainly where technology is used within small communities, or to provide for family members, or in a context of comradeship in the workplace. In other circumstances, it is often difficult to make a connection. Victor Frankl couples love and work as two intertwined ways of finding meaning in one's life, and sees it as a measure of the misdirection of technology that the deskilling and displacement of jobs by machines so often makes people seem dispensable. The individual worker often feels little sense of contributing creatively to a finished product, or connecting with the people who use it. The "consequent loss of meaning" is one of the great issues of our time.[4]

In the 1970s, visiting a factory making electric light bulbs, Howard Rosenbrock noticed that whereas some tasks had been automated, others were still being done manually. Wherever a machine was used to perform a task, the designer of the plant had taken trouble to ensure that full use was made of its capabilities. But where humans (mostly women) were still employed, each repeating a standardized task every 4.5 seconds, no interest was shown in their abilities. If the designer had even gone as far as "to consider people as though they were robots," he would have tried to provide them with less trivial work.[5]

This lack of interest in people relative to machines seems to reflect something of the outlook of individuals attracted to work in scientific and technical fields. Anne Roe's study of the psychology of scientists was discussed in Chapter 2, and it will be recalled that she found many more interested in things than in people and showing a habitual avoidance of some kinds of human concern. This was reflected in the tests and interviews Roe carried out, and although her sample did not include engineers, her findings could well apply to any who entered that profession via training in physical science.[6]

This is partly confirmed by a classic study of English schoolboys, in which Liam Hudson observed that many attracted into science were of a distinct personality type. He described them as "convergers," because they were good at solving the kind of problem that has only one correct, unambiguous answer on which their work could converge. This is exactly the type of problem common in school science, although it is untypical of ordinary life. Hudson also found that his convergers preferred to avoid situations with emotional or other complications. Often they were reserved individuals, and some were quite definitely loners, preferring to explore the impersonal world of scientific ideas rather than confront the uncertainties involved in dealing with people.[7] Significantly, neither Roe nor Hudson investigated the attitudes of those who *teach* science in schools.

Some writers on this theme, as we noted in Chapter 2, would categorize Anne Roe's scientists and Liam Hudson's convergent schoolboys as object-centered in outlook. Many people with this orientation perform best in highly specialized fields, and a majority who show this correlation between personality and interest in technology are men rather than women. One view is that women are "more balanced in their priorities," and less often object-centered in outlook.[8]

Some of these comments are quoted in a stark and disturbing chapter on the culture of technology at MIT—the Massachusetts Institute of Technology—in which Sherry Turkle remarks on the "severed connection between . . . People who are good at dealing with things and people who are good at dealing with people. " She adds that this "split in our culture" has many social costs, of which the first and most poignant "is paid by gifted adolescents."[9]

I suggest that another of the social costs is a lack of ethical awareness or moral imagination among some scientists and engineers, because ethics is primarily a people-centered concern, and cannot be adequately represented in terms of object-related categories. Langdon Winner comments that our immense technical expertise is coupled with "scandalous incompetence" in dealing with human questions. He suggests that this incompetence arises from "atrophy of the imagination" that seems to involve loss of awareness of human meaning, and leads us to assess technology only in terms of economics, efficiency, and risk. Efficiency may seem an appropriate criterion when it means "doing things right," but that is not always the same as "doing the right thing,"[10] which is the ethical question.

An object-centered approach also has difficulty over the intangibles with which ethics have to deal. Winner notes that a new technology may be banned if it can be shown to cause cancer in laboratory animals, but not if its use leads to some form of injustice, for example, to workers in a factory.[11] A cancer can be discussed in an object-centered way, but injustice is intrinsically a people-centered concept.

A further point to consider is what to do if one is aware (as I am) of an object-centered bias in one's own thinking. Sherry Turkle notes that many technology students at MIT are aware of the divide between their work and the real world of people, and are always on the lookout for ways of making connections. She reports that certain books are especially valued for the way they link technology to human concerns, mentioning as examples works by Robert Pirsig and Samuel Florman.[12]

These books do not use the terms "object-centered" and "people-centered," but address several related kinds of fragmentation and specialization, including what Pirsig refers to as a divide between "classic" and "romantic" outlooks. Others have characterized the problem in different ways, Martin Buber by contrasting "I-It" and "I-Thou" relationships, and Mumford significantly seeing the object-centered approach to technology in more active terms as *"power-centered,"* by comparison with the people-centered approach, which he represented as "life-centered."[13] Other writers identify the severed connection in technological civilization by pointing to further contrasts, talking about "the curse that severs work from play" and the "separation of man from nature, and of the growing child from the use . . . of his physical sensations."[14]

Contradictory Ideals

The tension between object-centered and people-centered attitudes may arise because technology as we now know it was arguably founded on contradictory ideals. On the one hand, there is real commitment to humanity and a genuine life-centered intention in the work of many engineers and applied scientists. On the other hand, there are enthusiasms and drives relating to powerful machines, visual pattern making, and also exploration, that have no direct connection with human concerns.

During the seventeenth century, when modern attitudes to science and technology began to emerge, Francis Bacon spoke with enthusiasm about "dominion over the universe" in almost the same breath as he advocated the ultimate people-centered project of merging science with "charity."[15] It was as if, with moral vision, these ideals of conquest and human concern could be harmonized. But we have yet to see it, and ethical conflict between these high purposes seems intrinsic to the practice of technology. There are also many questions to ask about the very concept of a people-centered science or technology. Does it simply indicate an intellectual acknowledgment that people are important, or does it demand a deeper commitment "to do justice and love mercy"? Is it narrowly anthropocentric, or does it take account of environmental factors? Is it possible to achieve a way of caring that informs everything a technologist does? Does it extend from cherishing nature to appreciation of people in their relationships with nature? Bacon's goal for science, and by implication, for technology, was that it should be "for the benefit and use of life," meaning primarily human life, but with the wording left sufficiently open for us to include sensitivity to all life forms. It is worth adding that although Bacon's talk of "dominion" makes him seem the arch-villain of early modern science to many feminists and ecologists, there is a considerable ambiguity in his writings, and we may find a positive aspect even while we take warning from his crudities.

One feminist reaction to some of the issues mentioned here takes up the suggestion that women or girls are less often convergers, or are less object-centered in their thinking than are men. One response in education has been to depart sometimes from the standard, object-centered textbook approach to attract more girls to the study of science or technology, because it is found that "girls have relatively wider social and humanitarian concerns." Some science teachers object that traditional methods were "rigorous" and that rigor is now being lost. But one can argue that better coverage of human aspects of technology can lead to deeper understanding of the subject, and is in the interest of both genders.[16]

Historically, it is striking that Bacon and others of his generation characterized the kind of science they favored as "masculine," because that is how they thought of factual, objective ways of thinking as compared with folk knowledge and intuitive thought. In some ways, they were taking much the same attitude as those who defend traditional rigor in science teaching and fear invasion of the curriculum by wider concerns. These attitudes may also reflect persistence of the idea of science as being concerned with control over nature, for Bacon quite deliberately wrote of masculine science in the same context as he claimed to be leading: "Nature with all her children to bind her to your service." Thus science and nature, with assigned gender roles, were supposed to benefit humankind.[17]

There have also been feminist reactions against the kind of ethics that looks for something like a Hippocratic Oath for technology. Abstract ethical principles such as this appeal mainly to men, it is said, when it is more practical and useful to think about ethics in terms of who is responsible to whom. Whereas men may be concerned with "caring about" some principle, women would prefer to stress "caring for." In their book on the ethics of engineering, Martin and Schinzinger acknowledge this in a limited way by quoting a comparison between (male) preoccupations with ethical rules and rights, and an "ethics of care" more concerned with relationships.[18]

The question then arises as to why gender enters so much into these issues, and hence into the human meaning of technology? Two reasons seem especially important. One relates to men's experience in society, which involves few responsibilities for children, the old, or the sick. The other reason, which we shall examine first, concerns work as a source of meaning, and the way basic production work was traditionally organized.

Gendered Division of Labor

In prehistoric times, it is commonly assumed, there was a division of labor in most human groups ensuring that hunting was a task carried out by men, at least with regard to the hunting of large animals as opposed to small ones such as rabbits. By contrast, the gathering of fruits, nuts, leafy materials, and roots was an aspect of production that seems to have remained a female concern, and that made women expert on plants and their uses, not only for food, but also for textile fibres and medicines.[19] Perhaps also women were mainly responsible for the domestication of plants and the beginnings of agriculture, and for inventing methods of preserving and processing foodstuffs.

Taking grain milling as an example of the latter, one cannot know much about how it originated, but it seems that from very early times, gathered grains and nuts were broken by pounding to make them eatable. Sometimes this was done by a hammering action using a stone on any available hard surface (which is what some apes and early hominids also did). Later, the pestle and mortar evolved. But when cereals were grown in quantity and processed separately from nuts, a more specialized tool was developed. A saucer-shaped hollow was made in a large stone and grain was broken in this by rubbing with a small, rounded, hand-held stone. An ancient Egyptian statuette shows a woman using such a device, and later, about 1000 B.C., there is evidence of the first hand mill with a round runner stone or quern turning above a bed stone.[20]

Hand-milled grain represented a domestic scale of production, supplying individual families. But the larger output obtainable when driving a bigger millstone with a primitive waterwheel (from 300 B.C.) was justified only if it served several families—that is, if the mill had a public or community function. And it is observed that many techniques change from being women's to men's work as they move from the private to the public spheres of life.

This way of reconstructing prehistory gives women an important role as inventors of pottery, textiles and food processing (including milling), but they probably worked with a non-specialist approach, informed by the whole context within which techniques were used and the overall process involved. So long as the hand mill was a piece of household equipment, it would be seen as just one part of the process of feeding and caring for a family. The pace of improvement would be slow because there would be no reason for devoting special effort to milling while all other aspects of domestic life, such as water carrying, cooking, fuel collection, and making clothes, were limited by primitive technology and low productivity. However, once mills were driven by waterwheels, milling was removed from the context of food preparation and family care and became the object of specialist expertise. Millwrights could build and improve mills focusing on the machines themselves and forgetting the overall process. They would bring to bear the more narrowly focused outlook that has arguably been characteristic of the male contribution to development in technology.

There are similar points to be made about spinning and weaving wool, cotton, and other textile fibers, processes usually carried out by women in early times that sometimes became male-gendered occupations when the equipment became more complex, as in some weaving looms or the complex spinning mules of the industrial revolution.

The transition from hunter-gatherer society to societies of farmers and artisans not only entailed change in technology—and gender relations—but affected religion also. Some of the writers quoted in the previous chapter who discussed deep ecology value the "nature religions" presumed to have been practiced by hunter-gatherer peoples. By contrast, Mumford argued that the world's "great religions" originated among artisans, farmers, and pastoralists, whose views of nature were rather different. Mumford included philosophy in his perspective, and mentioned Socrates, son of a stonemason, as well as Jesus the carpenter's son and Paul the tent-maker. He adds that Laotze (Lao Tzu), founder of Daoism (Taoism), had sympathies with craft workers, though he himself was probably a scribe.[21]

The implied distinction between *artisan religion* and *nature religion* can be further illustrated by comparing the first three chapters of this book, which mentioned several aspects of artisan experience, with Chapters 5 and 6, which described the outlook of some groups of hunter-gatherers. The latter mentioned people have a "sense of the sacred" in nature, and a strong sense of place, whereas the chapters on artisan experience discussed feeling for materials and aesthetic form. It might be expected, then, that artisan religion would use this experience, as indeed, some English Puritans did during the seventeenth century. Chapter 3 described how the latter were able to use the language of alchemy to relate artisan experience to their spirituality, for example, by speaking of "illumination" in both fields.

In the Bible, as Samuel Florman pointed out in writing about engineering, the Old Testament includes many passages that mention materials with artisan-like feeling. But the later records of Jesus of Nazareth do not find him speaking in the manner one would expect of a carpenter. In just one place, he is reported as saying that those who work with stone or "cleave wood" might expect to feel him with them.[22]

It could perhaps also be said that artisan religions are mostly patriarchal and reflect a characteristic gender typing of work. However, that is all rather speculative, and it is more useful to note analysis of agricultural mechanization in a twentieth-century Islamic county, where Ingrid Palmer observed that gender typing of tasks was liable to change whenever there were large increases in labor productivity. She argued that mechanization tends to strengthen patriarchy, mainly for reasons of economic relations within households. That is, men take up the work with highest earning power so they remain economically dominant, while women are left with subsistence tasks, or work with low earning power.[23]

Probably, though, economic motivations are reinforced by the symbolic meanings and personal satisfactions that men find in jobs that are technically interesting or involve powerful machines or some risk taking. Power and risk enter the equation where men feel their status to depend on being adventurous and "macho." In artisan societies, that may be a factor in a blacksmith's or miller's work, and in the nineteenth century, it accounts for the gusto with which men performed such hazardous jobs as manhandling white-hot iron (with tongs) in rolling mills, or riveting steel frames for buildings high above city streets. In Britain, women could occasionally be blacksmiths or bronze founders, or they might be employed in ironworks, but in the latter case, did only menial, low-paid laboring jobs.[24]

Again there are hints of the contradictory ideals on which technology is founded, for although many engineers and other practitioners are deeply concerned with minimizing risks to other people, and above all, with being of service, there has still been, until very recently, a feeling that some technological activities are so expressive of macho characteristics that they seem inappropriate for women. This was reflected in the reluctant, slightly awed admiration accorded to the likes of Emily Warren Roebling in relation to her work on the Brooklyn Bridge,[25] and Lady Charlotte Guest in her management role at the Dowlais Ironworks in Wales.[26] Similar attitudes were apparent even as recently as 1945 in the way women aircrew members were treated in the aftermath of the Second World War. During the war, women had flown transport planes for the British air force, and women pilots had delivered Spitfires and other fighters from the production line to the airfields where they were to be based. The women became enthralled by the adventure of flying and enthusiastic about their work, but when peace came, they were dropped by the air force, and once back in civilian life as housewives, some could not even talk about their war work to their husbands. The latter, presumably, would have felt that their masculinity was threatened by a wife who could fly Spitfires.

Such attitudes now seem bizarre (except in some churches where the idea of women controlling symbols of power is still not accepted). But in many of the examples quoted, technological context seems to have been almost as deeply involved as dress in defining the meaning of gender. A woman piloting a powerful military aircraft or Lady Charlotte Guest riding coal wagons around her ironworks could appear to be challenging convention in much the same way as if they were cross-dressing for a public occasion.

My understanding of this was helped by hearing actresses commenting on theatrical roles that involved impersonation of the opposite gender. They commented that women are usually more successful playing men than are men in an opposite role. This is because playing a man means exercising power; playing a woman means giving it up, which male actors find difficult. Playing a man is expressed in external action. Playing a woman comes from inside. A man often has a quest to do things in the world, and it can be exhilarating for a woman to experience this, even if just acting a part. In conventional drama, few women are "heroes," so few have quests.[27]

The actresses commented especially on the significance of footwear for different ways of moving, walking, and being, noting that feet can be especially significant symbols of power or submission. Boots express freedom and action, whereas fashions that restrict walking (high heels, tight skirts, and at one time in China, bound feet), emphasize separation from actively powerful life. They indicate a more seductive, if not entirely submissive outlook.

Tools and personal equipment provide means of displaying the messages conveyed by footwear on a larger canvas, as one sees with cars, dishwashers, office machines and leisure equipment, as mentioned in Chapter 4. Similarly in earlier times, hand mills and water mills, spinning wheels and looms carried similar messages about restricted domestic life or expansive and powerful skills.

Continuities between gender typing in work, dress, and consumer goods and gender typing in the practice of science are now easily recognized. Men more often than women have traditionally been attracted into engineering, nuclear physics, and other technologies dealing with powerful forces, although nuclear physics has seen some very distinguished women: Marie Sklodowska Curie, Irène Curie and Lise Meitner. But the tendency for these subjects to be dominated by men had been given much impetus when individuals associated with the earliest phase of modern science put forward the wholly artificial view that factual and practical knowledge is masculine (as we noted earlier in quoting Bacon). There is also the point that "distinction of sex roles is linked with . . . authority," and that science is seen as a source of intellectual authority.[28]

In the early decades of the nineteenth century, women could play a part in science by writing popularizations, or science books for children, but writing a university textbook in a mathematical field implied more authority. When she wrote such a book, Mary Somerville felt a considerable conflict between what was expected of her in terms of "womanliness" and her personal aspirations as a mathematician. At the end of the century Marie Curie's discoveries about radioactivity seem full of irony when we find that her husband, another physicist, had initially regarded science as an all-out masculine endeavor, hostile to "nature" as represented by women, and was amazed to find a soul mate in Marie.[29]

When punch cards were widely used to input data into computers, almost all those who prepared the cards (in Britain, at least) were women, because they had the keyboard skills. Some women also went on from this work to be skilled in programming, which initially seemed to develop as a gender-neutral occupation. Perhaps it has remained so in the United States where, in 1990, about half of all computer professionals were women. But in Britain, computing was often taught in schools as an adjunct to mathematics, which was perceived as a male-gendered subject.

As salaries rose, men became more competitively involved. For these and perhaps other reasons, by 1990, the proportion of British computer professionals who were women had fallen to 17 percent.[30] Similarly, with other new technologies, where one might have thought that traditional gender roles would be left behind, it again rarely turned out that way. In the development of microwave cookers, there was no real partnership between the women who were experts on food and male electronic engineers. According to one observer, the latter were dominant, and focused on equipment rather than on the processes of cooking.[31]

Women, Men, and Babies

Only one aspect of the division of labor in society has an inescapable biological basis, and that, of course, is the bearing and nursing of children. There is a common perception that this is one of the most meaningful experiences of life, and it therefore demands particular attention. For example, if men feel excluded from this area, does that help explain why they have sometimes appropriated other areas of life as exclusively their own? Does it show why, from the time of Francis Bacon, science and engineering have sometimes been characterized as the result of "masculine birth"?[32]

Similarly, if some women feel that raising children is a unique experience, does this affect their other interests? I have heard women speaking about bearing and bringing up children as the most intensely felt and fulfilling part of their lives. They talk about "motherhood altering their priorities." So do women engineers and computer professionals find their attitudes to technology changing after the birth of a child?[33]

The traditional view might have been that women's priorities in these areas are influenced by "maternal instinct." It might also have been thought that not only childbirth, but the whole process of looking after children, is a biologically determined aspect of the gendered division of labor. However, researchers have problems identifying such an instinct if it is understood as an urge to bear children. Apes and monkeys and other higher mammals are motivated to reproduce by the sex drive—by the urge to mate—not by an urge to bear young, as such. By analogy, a woman whose sex drives are satisfyingly fulfilled is not behaving in an "unnatural" way if she chooses not to have children, and she may be happier without them.[34]

On the other hand, where parents want children, there are good reasons why they are likely to find deep satisfactions in having them. As the opening paragraphs of this chapter suggest, contact with a child can somehow infect us with a fuller sense of what life is about than almost any other experience. Also, as some parents have said to me, a baby is "somebody else to fall in love with." Those of us who are childless may well feel drawn to find some of the same life-enhancing experiences by teaching other people's children, or even by keeping an animal that responds to us almost as if we were its parent.

A further reason for wanting children is the wish to feel that some part of us will go on into the future, after we are dead. That impulse can also be fulfilled by doing creative work that will last, as we shall see later. Every author hopes that her books will continue to be read after she is gone; every engineer wishes "to build a pyramid" by which he will be remembered (so one member of the profession once told me).

Biologists and psychologists also doubt the existence of a maternal instinct because they find that many parenting skills do not come to mothers easily or naturally, but have to be learned. Only a few behaviour patterns seem to be genetically determined. Efforts to identify inherited maternal behaviour in apes and monkeys have also found meager signs of it. Again, it seems that parenting skills are mainly learned, particularly from the mother's early experience during her own childhood. Indeed, some species of monkey, if deprived of mothering during infancy, do not know in later life how to care for their own babies and may abuse or kill them.[35] Humans who have had a deprived childhood also tend to make bad mothers.[36]

It also seems that some innate biological drives that do exist in humans with regard to care for children are equally shared by both sexes. In experiments that have measured people's physiological responses to babies, although teenage girls showed overt interest in a baby whereas boys pretended indifference, heartbeat and blood pressure responded in exactly the same way in boys and girls if the baby cried. So both were being alerted physiologically to help or comfort the baby. Similar responses are observed in adults, including childless people.[37]

It is also found that the first few days of a baby's life are important for bonding with the father as well as with the mother. In cultures where fathers are kept at a distance at the time of birth and for some time afterward, they rarely develop close relationships with children, and the view of child rearing as an exclusively female preserve becomes self-fulfilling. But when a father is present at the birth and helps care for the baby from the start, he can bond more closely with the child and can complement the mother in important ways.[38] Indeed, it seems that whereas most *skills* associated with motherhood (and fatherhood) have to be learned, both parents have a biological predisposition (or instinct) to care for children, and cherish them.

One further question that needs to be asked concerns the comparison sometimes made between creative work and childbearing. For example, it is occasionally suggested that the energy a parent pours into care for a family is no longer available for creative art or science. Once a woman has borne a child, it is said, her self-containment and individuality have been compromised. Until the child has grown up, a mother does not easily allow herself to be fully absorbed in creative work, or if she does, may feel guilty about it (as Kathleen Raine seems to have felt as she pursued her vocation as writer and poet, while a belief grew in her that women "should love first of all . . . love our children"). Artists, scientists, and inventors probably need to work in an object-centered way at least to some extent, whereas a parent has always to be person-centered. So when the burden of parenting is placed solely on women, they may well not feel free to take up other creative enterprises until that responsibility is discharged.[39]

The creativity involved in producing a book, an invention, or a picture commonly draws on the same well of feeling and imagination as caring for a child (which Susan Hill discovered) or for an elderly father (as I have found). Virginia Woolf recognized that childlessness left her free to channel energy into writing, but felt envious when she contemplated the fulfillment that her sister, Vanessa Bell, apparently enjoyed in family life. Yet Vanessa in turn had regrets about putting aside her art for the sake of her children.[40] Both sisters felt unfulfilled, though in different ways. But often a feeling of unfulfilment spurs creativity.

Love and work, some say, contribute more fully than any other aspects of life to one's sense of purpose and meaning, and most people need a balance of the two, not an enforced choice between one and the other. Where women's lives are concerned, Margaret Drabble used a story about writing a book and having a baby at the same time to suggest that a balance is possible.[41] The opposite extreme is the patriarchal society in which women were expected to find their satisfactions in the efficient performance of domestic routines. There may indeed be satisfactions in preparing food and keeping a house spotless, but anyone living this way inevitably feels that there is little to show for his or her efforts and wants to create something of enduring value as well. In some patriarchal societies of the last century, embroidery, dressmaking, or quilting were among the few approved outlets for women's creative energy, and women poured all their love of color and pattern, their natural energy, tension and ambition, into this restricted area. Even more recently, for women of the Old Order Amish in America, making quilts has been one of very few creative activities that were approved, and quilts have been "an unconscious expression of the need for individual achievement." Some quilts, indeed, "convey a strong feeling of tension," and express repressed needs and frustrations.[42]

By contrast, men may seem blissfully free from such restrictions, but the trap awaiting them is that work may become so compulsive and preoccupying that there is never much engagement with intimate, caring areas of life. Whole realms of meaning are then missed. This chapter began with a grandfather holding a baby and finding comfort in the baby's smile. That kind of experience, and its opposite, the bleak alienation of some men who have no such dimension to their lives, confirm my belief that men as much as women do not live complete lives if they distance themselves (or are separated by circumstances) from what they might learn through family life, or in a caring profession, or teaching, or even caring for animals. Of course, there are experiences that men cannot share, including the urge to protect a baby still in the womb, and the deep companionship felt with a child at the breast, and they often feel excluded. One artist has expressed this in a picture that shows a woman breast-feeding her baby, with other members of the family looking on.

The women in the group look serene and fully absorbed, but the father is a sad-looking figure who seems to feel that the women have access to a source of meaning he cannot reach. And whereas in ordinary life, women have often been required by fashion to dress as if their role were merely to be decorative, and not to be taken seriously, in this picture the man is dressed as if he had no serious function, for he is in the costume of a harlequin.[43] (Figure 3 on page 100 is another picture in this series.)

Among others who have felt this as a problem, one[44] cited an African ethnic group where a simple ceremony is held when a baby reaches the stage of beginning to take solid food. The father prepares the baby's first solid meal and from then on shares in feeding the child with its mother.

But in many other societies—and again oft-quoted examples are in Africa—the gendered division of labor operates to make men marginal to the whole life process, notably because women not only care for children, but do most of the agricultural work needed to provide food, as well as dealing with fuel and water supplies. In one community it was said that the only useful thing men ever did was to build houses.[45]

As to the responses of men who (usually subconsciously) feel themselves excluded from an area of life so filled with meaning, there is "a suspicion of false bravado" about several aspects of their behaviour,[46] such as their liking for jobs in which they can display boldness or strength, or their liking for army uniforms or business suits, which make them seem important. They may also look for compensatory sources of meaning, perhaps in a compulsive, workaholic lifestyle, or in very intense creativity in art or science, or in fanatical devotion to a sport or hobby. Alternatively, their alienation may have a more negative, more direct expression in drinking, drugs, or petty crime. Male culture may also sometimes express overt resentment of women, no doubt partly in reaction against their apparently more fulfilled lives. Such resentment has been well documented in a well-known study of male fantasies in interwar Germany, and in a journalist's account of attitudes to crimes against women in Britain.[47]

Writing more constructively about the alienation felt by many men, D. H. Lawrence thought they could escape it if they could learn to behave spontaneously in experiencing the full tender meaning of sexual relationships.[48] But he showed a certain lack of awareness when he ended the story of Lady Chatterley before her baby was born, and so before the experience of care for a child could demonstrate other kinds of tenderness and human meaning. It seems, indeed, that the alienated man often tends to focus excessively on sexual relations as the deepest source of personal meaning. If men would only stop thinking about sex, I have heard women say, they would discover that there are many other ways of loving.

Leo Tolstoy was perhaps more aware of this than Lawrence, and in carrying one of his stories through a birth, described the bewilderment of the young father, and his discovery of unsuspected dimensions of human meaning. Psychologists note that "immediately after delivery . . . mothers appear to be in a state of ecstacy," and observers become elated too.[49] The father in Tolstoy's account senses this through his emotional confusion, and notices the expression of bliss on his wife's face. His thoughts about "the meaning of a woman's life," which had only "dawned on him after his marriage," now almost overwhelm him. As for the baby, his feelings are "not at all what he had expected," but are dominated by an enormously strong "desire to protect."[50]

Yet even in 1998 it was possible for one woman who had borne a child to say: "I couldn't find a book to explain what was happening to me as I became a mother." She encountered "an amazing silence" about the experience she had of her whole outlook changing, as if nobody else had ever felt the same. Everyone seemed "to collude with the great white lie" that women could just have babies and go straight back to work as if nothing had changed for them.[51]

At the end of life, when bereavement breaks bonds with a sister, parent, child, partner, or close friend—even bonds that are not overtly affectionate—one not only feels grief and loss, but in addition, a change in orientation. After weeks and months have gone by and the worst pain has passed, one may still feel a different person. Even more, then, new bonds with a new person at the start of a baby's life can be a transforming experience.

The Gestation of Creative Work

Mumford's view of the great religions being founded by people of artisan background was quoted earlier to make the point that the creative work experienced by craftspeople can be a life-enhancing experience and a source of religious inspiration. But despite his experience as a carpenter, Jesus of Nazareth did not fit this pattern very well, we noted. A better way of putting his teaching into context may be to think less of his artisan background, and more about the kind of experience just quoted. He was close to a number of women, including his mother, and this could be the clue to some of the striking things he said about love. "The loving relationship of parents with a small child has a numinous quality which the fortunate child can carry" into adult life, according to one psychologist.[52]

The Greek word *agape* used in reports of what Jesus said about love may refer to this; and it can also be taken to mean something close to "cherishing," as one cherishes a partner, child, or friend (and as the word was used in chapter 5 to mean both valuing and caring for). But the translation Francis Bacon used when writing about science rendered the meaning of *agape* as "charity." Thus he said that those who seek knowledge of a scientific (or technological) kind should "perfect and govern it in charity,"[53] deliberately using Biblical language.He warned in the same passage about "lust of power," adding that by contrast with excesses in that area, "of charity there can be no excess." Yet we noticed earlier how Bacon also asserted that the proper role of science was to give "man" dominion and power over nature.

Bacon's writings influenced ideas about the practical uses of science throughout the seventeenth and eighteenth centuries, mainly in Britain, but to some extent in France and America as well, and as noted earlier he may be said to have set out conflicting ideals. No doubt dominion over nature was regarded as a means to the end of using science in charity, but it could easily become an end in itself, distracting people from works of charity. But perhaps one should not blame Bacon too much for the contradiction, because it seems to be a tension that has always been present, extending even to concepts of God. Thus on the one hand, some people seem to build their ideas of an "omnipotent" God around their fantasies of ultimate power, dominion, and control, including the handful of modern scientists who enjoy speculating about the possibility of finding God in (or beyond) the Big Bang that supposedly inaugurated the universe.[54] On the other hand, there is the much more stimulating notion of God as a metaphor for love (without power). This asserts not only that "God is love," but adds, "let us love one another: for love is of God: and every one that loveth . . . knoweth God."[55]

However we choose to resolve the contradictory impulses underlying inherited ideas in this area, the varieties of love, whether sexually driven, evoked by a baby, or stemming from friendship, together provide the most important reasons most people find for living. But it is also striking that some people bring their experience of creativity into the equation.

Elizabeth Goudge, a novelist whose work is no longer in fashion, implied in an autobiography that her books were a substitute for the children she never had. At the same time, she made a link between writing and other craft skills. Thus carpentry was based on "love of the wood . . . and the slow labour of craftsmanship," and in writing a book, there has to be a similar love of the work. In that and in having a baby too, she said, there is also a "glowing imagination."[56] In other women's lives, writing a book and having a baby occasionally coincide. This happened in 1749 to Emilie du Châtelet when working on her great translation of and commentary on Newton's *Principia Mathematica*. In an age when many women died in childbirth, she greatly feared the consequence of her late pregnancy (she was past forty). Thus she redoubled efforts on the book to ensure its completion, and she was at her desk when labor began. The baby came quickly into a book-strewn room and "was laid on a quarto tome of geometry." Emilie's forebodings were sadly justified, and a few days later, she dated her completed manuscript and, shortly after, died.[57]

There was a feeling here that the book was an infant that had to be successfully brought into the world as well as the real child, and Marie Curie expressed the same thought as she continued laboratory work during her second pregnancy. She cherished her first child, but confided to a woman student that "radioactivity was also the child to which she had given birth," and she cherished that too. But she was not fully aware of the dangers of her scientific offspring, and in the summer of 1903 was unwell with "pregnancy sickness combined with radiation sickness," ultimately losing the baby.[58]

Worlds without People

Although there are intriguing parallels between creativity and childbearing, there is a sharp contrast to be drawn in some instances where people create without the glowing imagination of which Elizabeth Goudge spoke, but rather with a wish to control. One thinks of individuals—more often men than women—who enjoy bringing something into being through a process of design and calculation in which nothing is left to chance. This is the opposite of the experience of most artists, inventors, and scientists that their work grows beyond anything they could have planned or intended. Inventors, in particular, are sometimes surprised by how their creations are ultimately used (as Edison was with the phonograph). Composers and dramatists also have to allow their work to have a life of its own in the hands of its performers. Parents have to allow daughters and sons to develop their own individuality beyond anything they can plan, and often differently from what they hoped for.

But there are those who wish to design and calculate so that what they create fits closely with the original blueprint and brings no surprises. This kind of pleasure in control can also be expressed by interest in automatic systems perceived as little self-contained worlds without people. Men who (like me) are fascinated by model railways may be playing with a form of this idea, because each such system is visibly a little world in itself. Significantly, the chairman of a German company that manufactures model trains sees a desire to control things as a reason for their popularity. It is said that computer games have now partly taken the place of the train set; this may be because they offer enhanced opportunities for control of little worlds without people. The same wish to control without the unpredictability of real humans may also lead to interest in robots, automata, and computer programs, all operated as "quasi people." And of course there are often attempts to exercise a will to control over real people, often disastrously.

David Noble and others have speculated that what drives men in particular in pursuing creativity related to control may be an "urge to compensate for lack of the female capability of giving birth to children." Some also say that the attitudes of men who work in obstetrics, gynecology, and the new reproductive technologies reflect an urge to take control of an area of life from which they are otherwise excluded.[59] Building robots "might be seen as the ideal compensation," and Noble adds that interest in devising automata has a long history as a "peculiarly male preoccupation." An individual automatic mechanism "satisfies an enchantment with things that are at once animated and artificial."[60] David Lodge presented another aspect in a lighthearted novel that describes a computer numerically controlled machine tool working automatically on cylinder heads for car engines. At times, there is an uncanny sense that the machine might be alive, "like some steely reptile devouring its prey." A factory full of such machines working in coordination with one another would be the ultimate world without people, and we should consider how this has become a paradigm or model for much thinking in technology.[61]

The most famous account of this male impulse to construct automata and other artificial life is Mary Shelley's novel *Frankenstein,* written in 1816–17 at a time when the scientific fashion was not computers but galvanism (including electric treatments for disease and experiments in which corpses were made to twitch by subjecting them to electric stimulation). Thus the fictional Victor Frankenstein assembled a body from materials furnished by the "dissecting room and slaughter-house," and was able to infuse an electric "spark of being" into it.[62]

As a caricature of the motives and behaviour of the scientist (and today, the technologist), the story is very much on target. Langdon Winner described Mary Shelley's book as the closest we have to a definitive parable of our "ambiguous relationship" to technological creation.[63] If one asks how a nineteen-year-old author could have gained insight to write such a book, one answer is that both her parents had written about themes she took up; another is that her husband, Percy Shelley, had speculated about reanimating corpses by galvanism. But a more serious model was provided by Humphry Davy, the chemist, whose writing Mary had studied. Her book, indeed, can be taken as a reply to the printed text of Davy's inaugural lecture at the Royal Institution in London, delivered in 1802, in which he claimed creative powers for the scientist, enabling him "to modify and change the beings surrounding him."[64]

With considerable outrage, Mary caricatured this kind of scientist "to a wry perfection," picturing him in her description of Frankenstein as endowed with power "to reproduce without women."[65] That brings us to another reason why Mary Shelley wrote so well on this theme. She had previously borne two children, and she completed the book in May 1817 when pregnant with a third. As some critics point out, what Mary did was to express her experience of pregnancy and childbirth by transferring it to a male actor—but a male actor who wanted to control rather than love.

Other interpretations of Mary Shelley's *Frankenstein* have been proposed, some stressing her brief comments on alchemical writers,[66] and many noting that the women in the book, though idealized as carers, are all shadowy figures. This may be because the story demonstrates a male desire to take control of all significant roles, reproduction included, leaving few other functions for women to perform. In other ways, the book can be seen as prophetic, anticipating the enthusiasms of later generations with opportunities to work on robots, artificial intelligence, genetic manipulation, and the new reproductive technologies.

However speculative and tentative one must necessarily be in covering so broad a topic, the effort to recognize connections is at least fruitful in raising questions, not all of which can be discussed here. Two themes, though, seem especially relevant: First, what would technology be like if it had grown from ideals of love rather than of power, from people-centered caring, and traditional women's roles, instead of from the other half of the old division of labor? This matter is discussed in Chapter 9. The second issue is in some ways the converse of the first: What happens when technology is employed by individuals who feel excluded from the most meaningful experiences of life and alienated from family relationships? Or when it is exercised by men with attitudes like Victor Frankenstein's? We discuss some answers in Chapter 8.

A third issue concerns the seemingly inescapable need in a heavily populated world for many people to adopt a lifestyle in which raising children plays a smaller part. Many women in the industrialized countries are already opting not to have children. Although this is a welcome development in many ways, it raises all sorts of issues about an aging population with fewer younger people. But we can surely now appreciate another point, one about how people find meaning in their lives. Having a family was traditionally the culmination of many people's life experience. Other kinds of loving, cherishing, or caring for others may provide a comparable sense of meaning, and so we have seen, may creative work.

However, if there is less opportunity for people to find fulfilment in love of family, there may be greater need to show love in other aspects of life—to love beyond the family. For example, a correspondent writes about his efforts to "let love inform my interpretation . . . of the practice of management." That kind of language is difficult for most of us because of the connotations of intimacy which usually go with the word "love," and in many cases, the word "cherishing" may be more helpful. For myself, I *value* the students I teach especially for their commitment; I *empathize* quite strongly in the problems they encounter; and I do my best to *care for* them within the limits of academic responsibility. All three points fit the definition of *cherishing* quoted in Chapter 5, and that term also fits most relationships with colleagues, friends, and neighbors. So even though close friendship and intimate affection include only a handful of people, I can extend some aspects of love to a great many.

Moreover, that wider, cherishing, empathizing kind of love can inform work in technology, creating opportunities for conviviality in factories and workshops and fostering a responsive sense of service to the community. Not only that, but the implication of this and the two preceding chapters is that the idea of *cherishing* has potential to replace the aspiration to *control* as an ethically preferable paradigm—or model, or ideal—for what technology seeks to do in relation to nature, as well as with respect to people. In other words, there should be a much greater emphasis in the practice of technology on more adequately valuing and caring for both people and nature.

We will explore this latter possibility further in the final chapter, but before that we must consider some consequences of the opposite tendency, represented notably by object-centered styles of work in technology and by aspirations to create worlds without people.

8

Knowledge Pregnant with Evil

Values in Technology and Science

It is often claimed that scientific research is so rigorously detached, and so carefully isolated from personal feeling, that science is neutral with regard to ethical values, political affiliation, and all purposes apart from the one purpose of seeking knowledge. Yet when scientists are told that some experiment they wish to perform may not be permitted—an experiment on cloning human cells, perhaps—they tend to protest, not only in the name of science, but also in the name of progress. For many people, indeed, technology and science are not at all neutral but constitute a movement for good in human affairs. Even if Bacon's hopes for science being practiced with "charity" and "for the benefit and use of life" (discussed in the previous chapter) have sometimes been denied or subverted, a widespread view still identifies science and technology with positive benefit.

If we think of the individual practitioner rather than general trends and assumptions, we must certainly notice that he or she can rarely be value-neutral in attitude. Chapters 2 and 6 noted that some scientists and engineers, like explorers in the past, have an object-centered orientation and appear to lack interest in other human beings. Chapter 4 noted that for a few, research can be an extension of childhood play into adult life. The collecting and classifying activity in some sciences may exemplify this. So may the interest in investigating how things work displayed by boys whose play includes dismembering small animals. My own childhood play often took the form of projects I devised for myself to be carried out during school holidays, and I can see a clear continuity linking this to work I did as a scientist during my early adult years, as well as to the work I do now, writing books or researching the history of technology. One teacher and mentor of my student years openly admitted: "My devotion is to projects," whereas his wife's was "to persons."

Even if we make allowances for the maturing of an individual's interests and the practical requirements that discipline the work of engineers more than scientists, each individual is likely to carry some bias of these kinds from early years. In each of these examples, the wish to be in control of a project or a collection or other entity is likely to be part of the motivation.

There are other characteristic values in science and technology as well. Some people undertake experiments with care and restraint, especially when studying other living things. Others, though, are strongly interventionist in their way of practicing science, and would never wish to shirk the dangerous experiment or drastic vivisection.

In ordinary circumstances, these divergent attitudes are not great, and we can think of science as a cooperative enterprise whose practitioners work together with widely shared values. However, there is potential for much greater differences to emerge under extreme social or political pressures. The classifiers and control freaks may find greater opportunities for developing their distinctive approaches in a society addicted to bureaucracy. The individual who spent his boyhood pulling plants or animals apart, or who prefers interventionist forms of scientific experiment, may feel that he is "given permission" to develop these tendencies further under extreme conditions (such as warfare), or when government authorities behave ruthlessly. Much technology has been "conceived and applied in the context of war and oppression," yet many still want to think of it as morally neutral, as if it bore no mark of its origins.

Although it is admirable if practitioners of technology and science strive for neutrality and objectivity, this should be seen as an ideal that is often unattainable. What is more necessary is awareness of bias and influence. We also need precautions against the political and individual ruthlessness that have occasionally combined in such a way that the practice of science becomes pregnant with evil, as we shall see later in this chapter. Science, at its best, is the highest achievement of the human intellect, but the practice of science and its application in technology is full of paradox in which the noblest ideals of truth (and Baconian charity) coexist at close quarters—sometimes in the same minds—with tendencies strongly inimical to human life.

The paradox is clearly evident in almost any community where military research or arms manufacture is a major part of the local economy. Debra Rosenthal's comments on Los Alamos as a community make the point well,[1] although I also have in mind towns in Britain where the manufacture of military aircraft is a major form of employment. In their day-to-day life, these are thoroughly decent communities whose ethos is compassionate and progressive. It is therefore hard not to think of local business and community leaders as people of goodwill and integrity. If they say that the work being done in local factories or laboratories is a matter of necessity or duty, and subject to all kinds of safeguards, the ordinary citizen usually accepts the reassurance. Newspapers may document the export of military equipment from a local factory to some nation involved in aggression in the Middle East, or its use in genocidal activities against the people of East Timor (to cite one particularly scandalous specific instance). But people still soak up the ready-made assurances they are offered, and when their government claims that arms exports are subject to an "ethical foreign policy," consciences are further eased. In both Britain and the United States, the kind of ethical policy most urgently needed is one that would first reduce and then eliminate the dependence of swathes of industry on large-scale armaments manufacture.

In asking questions of workers, engineers, and scientists employed in "defense" industries, one hears a wide range of views about the development and production of weapons. Some people have thought carefully about their country's defense and export policies and have decided to support them. However, most people are less deliberate, and may explain their involvement as due to limited career prospects or income-earning opportunities elsewhere. In addition, when scientists and engineers work away at specialized problems, they may be thinking of the overall progress of technology without considering specific applications. Indeed, applications may seem very distant from the analytical and technical tasks that are the everyday reality of the job. Sometimes, too, a technical problem may be so challenging that it would seem an evasion not to tackle it, and individuals may become fascinated by the puzzle-solving aspect. Or researchers may have a compartmentalized way of thinking, and so put aside, even forget, anything they know about the negative aspects of the technology they work on.

A minority of individuals in most places with this kind of industry, however, appear enthusiastic about the development of new weapons or faster aircraft, as they may also value more drastic ways of exploiting natural resources, because such things symbolize enhanced human mastery and power. Indeed, as we shall see later, there seem to be people whose mind-set is decisively oriented to technologies expressing power, however violently. Some individuals who think like this may be of the type who tend to assume "I'm different; the rules don't apply to me." They imply that they can afford to take greater risks than most people because they have better command of their situation. Although only a handful of scientists and engineers are like this (though perhaps more politicians and members of the military), they seem often to become project leaders, and their influence can often be detected in the pronouncements that come out of the nuclear industry, or from agro-industrial laboratories involved in genetic engineering.

Readers who prefer political analysis to discussion of mind-sets may recognize here the basis of two kinds of relationship between technologists and those who make decisions and form policy, whether in industry, the military, or government. On the one hand, scientists or engineers who pursue strictly technical research, or work compulsively without thinking about consequences, are vulnerable to being manipulated and used, and may become what some have called "servants of power," ready to be reassured by the locally-dominant hegemony.[2] Ideologies about progress in science and technology may even mislead them so that they fail to perceive the real objectives for which they are working. By contrast, those who can find satisfaction in violent uses of technology for the sake of what they symbolize about power have been known to promote weapons, or other destructive innovations, before there is any demand for them from the military. They have also lobbied decision makers to improve the prospects of the innovations they favor. This has sometimes been described as a "technology-push model" of the relationship between technologists and decision makers.[3]

Problems of Lack of Awareness

In seeking to understand the lack of awareness evident among many who work on weapons, we need first to recognize that many technological tasks can be enjoyable and absorbing. They can seem wholly positive and fulfilling, as the early chapters of this book aimed to show. That is most clearly seen in the older craft technologies, as when a metalworker spent many hours fashioning a finely crafted sword or gun.

In the most simple practical tasks, there are moments when concentrated, object-centered attention has to be given to the manipulation of materials. At such times, the wider purposes of the task are forgotten. Forging the steel for a sword by traditional methods was a long, demanding process requiring great concentration.[4] Then when the sword-smith could take time to think about the ultimate purpose of the task, he would probably picture the fine-looking object he aimed to create rather than the scene of a battle with mutilated bodies. And he would plan a decorative finish for the sword to celebrate his own skill and the owner's status—also, probably, to symbolize the "heroism" and "glory" of war. In other words, this was a task where switching to people-centered concerns would first be inhibited by the demands of the work, and then would be selectively positive, often regarding war as "glorious," when at the very best it can be only a tragic necessity.

Similarly, in modern laboratories, individuals responsible for the development of new types of nuclear bomb have sometimes felt that, in technical terms, they are pursuing "subtle, beautiful and innovative designs," rather than planning for death and destruction.[5] However, this view was not always easy if one was working on the actual functioning of a bomb rather than the more basic technology that went into it. At the Los Alamos and Sandia weapons laboratories in New Mexico, Debra Rosenthal encountered several people who were prepared to do limited technical jobs connected with the production of nuclear weapons but who would not take part in the design and manufacture of bombs because that would be "too close to pulling the trigger."[6] They would then feel compelled to think about the end result.

In such instances, there are habitual ways of thinking that focus on materials and technical efficiency and distance the individual from any experience related to what the technology might do to people. Some individuals at Los Alamos needed this distance; others could feel that it was the right policy choice for their country to build nuclear weapons. But most had some moral qualms, and one way of coping was to keep people-related concerns in a different mental compartment from any thinking about work. For some scientists and engineers, object-centered, compartmentalized thinking became such an ingrained habit that they seemed to be entirely out of touch with feeling for people. Rosenthal's most telling comment concerns the experience of marriage counselors at Los Alamos. They regularly encountered scientists who thought it sufficient to analyze intellectually the problems that were making their wives so angry. The counselors had to challenge this way of planning to "fix" every problem, and had to begin from basics "to explain empathy to these men."[7] Similar stories are told about men involved in other highly challenging and technically absorbing projects, including work for computer firms and the NASA space program. People-centered, empathic approaches get almost permanently switched off.

Development of the division of labor and its elaboration in bureaucracies can also make it more difficult to see the connections between one's day-to-day work and the people affected by it, as can the operation of a market economy. Wherever work is highly specialized and bureaucratic, or remote from the ultimate purchaser, the individual is usually aware of only the immediate task. Technical responsibility in the office or factory and financial rectitude in the marketplace become more important than moral responsibility for what products or processes actually do.

For example, subdivision of tasks in a napalm factory, and then the fact that nobody involved in production is connected with its military use, ensure that no individual is wholly responsible for the burned babies that result. Most workers in the factory are likely to perceive the process they operate as a morally neutral application of technology, and the possibility that any human being might be burned would be unintended by them. When napalm was first used, it was not always effective because, if the victim was quick enough, he or she could scrape it off. During the Vietnam War, industrial chemists were asked how to make it stick, a request that could be dressed up in scientific abstractions and that was dealt with by adding polystyrene. For the chemists, this was a narrowly technical question, and they did not have to dwell on the way napalm would now "keep on burning right down to the bone." Nor did the aircrew who dropped it on Vietnamese villagers have to think about this as they had fleeting glimpses of their targets. Nobody came face to face with the reality of what they were doing.[8]

People can in these ways be creative in tasks with unthinkable objectives and often wish to excel in the technical performance of what they do, because bureaucracies and the division of labor have the effect of turning means into ends. Routines in laboratories, factories, and workshops, or in flying aircraft, become jobs to be done conscientiously, as purposes in themselves.

This even applies to much of what happened during the Holocaust in Nazi Germany—the systematic killing of the Jewish population of all territories the Nazis invaded. What makes this especially dreadful to contemplate is that it was not the result of mob violence, mass hysteria, or a sudden impulse, but was the action of many people acting deliberately in a modern bureaucracy to carry out a "scientific" plan. As Zygmunt Bauman explained it in his classic study, once means became ends, extermination became a routine, in which efficiency of performance was the goal.[9]

Although this does not explain everything that happened, Robert Jay Lifton[10] agrees that genocide was made possible by the way bureaucracy could disguise the purpose of much of what went on by means of its paperwork and routine, and by the use of obscuring language such as "resettlement" to mean transportation to a death camp, and "special treatment" to mean killing. People who were daily doing the most dreadful things were able to regard them as unavoidable "duty" to the system and in support of colleagues.[11]

Beyond that, though, the Nazi regime cynically exploited the habitual lack of awareness of many technical people. Hitler's architect, Albert Speer, commented on "the technician's often blind devotion to his task," saying that many technicians could be led to work "without any scruples."[12] According to Bauman, this was true of the designers of the gas chambers and "gas vans" who could focus down, in a perverse manner, on technically interesting detail. The equipment was first used for killing the "insane" and "disabled" in a forced euthanasia program before it was applied on a larger scale for killing Jews. Gas vans employed carbon monoxide from their engines as the killing agent, and their designers developed much obscuring, euphemistic terminology to describe the work, removing all human reference. The people being killed were described in object-centered language as the van's "cargo," to be assessed only in terms of weight, fluid emitted (when they vomited), and gas intake. Makers of the gas vans used in early killings reacted with alacrity to complaints from operators of the vans when these reflected on their technical competence—but never questioned what the vans were for.[13] Similarly, makers of military aircraft in Britain react positively to technical complaints from oppressive regimes in Asia that use their planes, but feel it is not their responsibility when evidence is produced that the planes are used against ethnic minorities or impoverished villagers. Given these attitudes, and the use of object-centered language, it was—and is—possible to perform the technical analysis required, and perhaps even enjoy it, without facing the horror of what is being done.

Technological Distancing

The division of labor within a bureaucracy, or the chain of transactions in a free-market system, has the effect of distancing the individuals concerned from the end result so that no one person need feel especially responsible. In a more literal and physical sense, one of the main themes in the technological development of weapons throughout history has been to create distance between the user of the weapon and the victim. So progress has "consisted mostly in eliminating . . . the chance of face-to-face combat."[14] The invention of artillery that could hit a target invisible to the gunners was a great step toward achieving this, as was use of computer screens in bomb aiming so that use of the weapons seems even more remote from real damage to real people. As Zygmunt Bauman said: "Moral conflict does not arise from pressing a button." He added that in the circumstances of modern warfare, information technology "more than any technology that preceded it has succeeded in obliterating the humanity of its human objects."[15]

In issuing a similar warning, Robert Jay Lifton has reflected on the effects of technological distancing experienced during the Vietnam War as recounted by veterans he interviewed. In bombing from high altitude, the B-52 aircrews pressed a button but saw almost nothing of what happened, and they could speak of their work with professional detachment. But the crews of helicopter gunships saw the people below them quite clearly and knew the damage their weapons did—the people they killed. So it was they who came back asking questions, and feeling horror and guilt.[16] And Lifton went on to note that nuclear weapons are controlled by people in the more detached, awareness-lacking kind of situation.

The converse of such situations existed until recently among some peoples in Africa, the Americas, and Southeast Asia who still had no modern weapons. Whereas some had a culture of violence that sanctioned brutally bloody forms of warfare in which many people were killed in direct, personal combat, others had ritualized their conflicts so that very few deaths occurred. Among some nomadic peoples, including perhaps the Turkana in Kenya, battles might be staged about such issues as territory or theft of cattle, and fighting might briefly be fierce. But if somebody was killed or badly injured, such feelings of horror were aroused that the fighting immediately stopped, the dispute was settled by negotiation and an exchange of cattle, and the peace was sealed by a ceremonial burying of spears. But once the Turkana acquired guns, and could kill at a distance, these customs ceased. The sense of horror that had previously stopped wars was no longer experienced because death was not confronted at close quarters. The result was that fighting became far more devastating and lethal.[17]

However, instances like this where the distancing effect of modern weapons has led directly to increased killing do not explain many of the massacres in human history. There are numerous examples of people butchered in hand-to-hand fighting or killed in cold blood. Nonetheless, it is normal for human beings to feel horror at such things, and to be traumatized if they become involved, or merely witness them. This is true even of tough-minded fanatical men who seem to have put all scruple aside. Among troops specially selected by the Nazis for the systematic killing of Jews by shooting, there were many breakdowns and some suicides[18] in response to the scenes of horror that ensued. The men felt acute distress, especially when the groups being shot included children.

Yet none of this sufficed to stop the shootings, not even when the senior general who supervised them, Bach-Zelewski, himself had a serious breakdown with bad hallucinations: flashbacks to scenes of murder.[19] Such experiences instead encouraged the Nazis to develop the gas chamber as a form of killing that the perpetrators did not have to witness directly. The gas chamber was a means of technological distancing, therefore, and one man who had "shuddered at . . . carrying out exterminations by shooting" expressed relief to think that with the gas chamber "we were to be spared all these blood baths."[20]

Two points emerge about technological distancing. One is that where cultural values foster violence, or affairs are driven by perverted ideology, and where there are authoritative orders issuing from a bureaucratic command structure, killers will do their work anyway; distancing just makes it psychologically less stressful for them. The other is that although technological distancing can make the operators of equipment less immediately aware of the consequences of their actions, it does not account for lack of awareness among researchers, designers of equipment, or engineers. To understand why there is an awareness problem with these people—a lack of moral imagination—such that the Nazis could think that many technicians were "blind," we need to look elsewhere, perhaps to wonder whether these are further problems due to compartmentalized thinking. But another, possibly more potent reason, is the especially absorbing nature of the work done by many researchers and engineers.

Compulsive Technology and the Frontier Mentality

One way of understanding the compelling, absorbing character of work in some kinds of science and invention and how it can make some people "blind" is to think of a trivial comparison, namely, the experience of those who solve crossword puzzles or play chess and know what it is to encounter problems that, for a short time, become all-absorbing. One may spend far longer on a game than intended, or become unreasonably resentful of interruptions. Something about puzzle solving, and certain other challenges also, encourages an especially engaged and focused kind of thought. Such concentration is so focused, indeed, that it presupposes compartmentalized thinking for the short time during which the puzzle is being dealt with.

But this response of the individual to the challenges of puzzle solving interlocks with something far more dynamic, and extending through whole communities, that may be encountered when we hear technology spoken of as a "great adventure" worthwhile for its own sake, and when we hear commentators locating innovation on "something called a frontier."[21]

The frontier of settlement in North America, as it moved westward from the seventeenth to the nineteenth centuries, was a place where challenges of many kinds could be overcome and where there was opportunity to start a new life. But when the continent no longer had any wilderness areas suitable for settlement, people identified the "next great frontier" as technological achievement of certain kinds. Thomas P. Hughes has commented that the century between 1870 and 1970 was a period of unqualified "technological enthusiasm" in the United States, when individual inventors were very prolific, and when there was also a new awareness of the potential for devising and organizing large technological systems for such purposes as electricity generation and the manufacture of automobiles. Rather than stimulation from the challenges of a geographical frontier, then, there was a new kind of "creative spirit manifesting itself as the building of a human-made world."[22]

The rhetoric of the frontier could still be used, though, as it was most famously in the 1940s by Vannevar Bush in a report on science policy in the United States entitled *Science, The Endless Frontier*.[23] Winner suggests that the frontier, with its enduring "power over the imagination," has given a distinctive purpose and direction to American work in many of the sciences. Today, funding and political support are still attracted to projects that seem to have frontier characteristics: space exploration, the human genome project, and even, for a while in the 1980s, the "high frontier" of the Strategic Defense Initiative (the Star Wars project).

The prolific invention generated in all these areas depended on individuals with a compulsive, adventuring style of work, which can seem unbalanced and disturbing, as it seemed to Mary Shelley in her portrait of Victor Frankenstein as the dangerously driven man of science. But whereas some see Frankenstein as a cruel caricature of real scientists, others see his kind of drive behind much creativity, and in many triumphs of rational thought. For those who take a more negative view, what seems wrong, usually, is the sense that responsibility for future change is allowed to give way to the momentum of innovation. This seems just to carry us along, and it is as if "we cannot stop inventing because we are riding a tiger."[24]

That mood prevailed during the development of the first atomic bomb, when "it would have been contrary to the spirit of modern science and technology to refrain voluntarily . . . from . . . a new field of research, however dangerous it might be for the future."[25] Similarly, in exploiting the gas and oil resources of a "northern frontier" in the Canadian Arctic, "We . . . find it difficult to resist technological challenge."[26] All these situations display evidence of a strong impulse to say that if anything can be done, then it should be done, a compulsion sometimes referred to as the "technological imperative."

This attitude, widely accepted, influences policies for the organization of research as well as for the development of such things as bombs and pipelines. It reflects compulsive work by individuals and group responses to technological frontiers. It also reflects the momentum of large institutions committed to major projects. In many respects, the technological imperative needs to be understood as a social and political question. But we also need to recognize how it is related to the experience of individuals caught up in challenging work. For example, David Noble has commented that people seem driven to invention in almost the same way as they are driven to climb mountains, "for reasons no one has ever clearly expressed." But earlier in the same passage, Noble did express very clearly what it feels like to be driven this way, describing how one gets "emotionally involved when trying to make something work, whether the challenge is manual or intellectual. You skip dinner, ignore the calls of nature and other people, push on into the wee hours, driven, possessed, determined. There is delight in it, a passion—and a blindness." Yet "such emotional enthusiasm is the wellspring of creativity and can often be inspiring and enriching."[27] Indeed, one psychologist has suggested that "the ability to channel one's interests, even obsessively, may be a condition for producing original work."[28]

More recently, psychological studies of this state of mind have focused on people who spend long hours alone working with their computers.

One group of authors, indeed, has characterized computing as "compulsive technology," especially in the hands of such people.[29] The point is made that although some people with scientific interests throughout the generations have preferred solitary work, "never before has there been an activity . . . which could give the distinct impression of companionship and . . . intellectual stimulation" as the machine reflects back or develops the information fed into it.[30] Another view is that computer programmers and hackers are "like poets and artists, possessed by their media." Their obsession is not with the computer itself but with the issue of control over its operation, and what creative work they can do with it.[31] Hackers argue that they do it purely to meet a challenge, not for any information they may gain.

But although some creative ideas come from lone scientists who work in this way, it is more significant today how individuals working together in a laboratory or workshop may be mutually challenged by the intricacies of puzzle solving in a technologically purposeful environment. The commitment and enthusiasm of different people then becomes interactive and self-reinforcing, and very often, a synergy develops between individuals' projects and institutional goals. Tracy Kidder has memorably described the way this happened during the development of a new computer at the end of the 1970s, noting how the leader of the project detached his team from the rest of their company so that they formed a tight little community. Then team members tacitly "agreed to forsake family, hobbies and friends" for the duration of the project.[32] There was considerable dedication, if not obsessiveness, in all this, and a sense of conviviality and enthusiasm that can be greatly admired. But accounts of a similar style of work in weapons research are more worrying, and in other computer projects there is an impression that the people involved are not real enthusiasts so much as the "serfs" or coolies of computer culture. For some people, the experience is deeply alienating, and while the enthusiasts become "withdrawn and humourless" as the work becomes more absorbing, a few have lost girlfriends or found family tensions mounting.[33]

The compulsive style of inventive work or research considered so far is most often a response to challenges, especially of a frontier kind or involving puzzle solving. It is a response, moreover, in which some people are aroused to an enhanced level of alertness, activity, and competitiveness. Another, more widespread form of arousal, affecting the body as well as the mind, is the thrill which many people feel when they have control of a physically powerful device, especially if it is a motorcycle or gun or some machine that also sounds powerful (Chapter 1).

Car advertisements regularly appeal to such feelings, as we noted in Chapter 4, and in some markets, they may explicitly stress aspects of a vehicle's styling and noise that express engine power. Many drivers, of course, are more interested in a practical, domestic-looking car, but for those drawn to a powerful machine, the pleasure of driving arises partly from physiological arousal. An increase in pulse rate related to a hormonal response has been observed in some drivers each time they accelerate. But this arousal seems to differ from the enhanced alertness of somebody responding to a challenge. It may be more aptly compared with the feeling of control that some computer enthusiasts enjoy, because of the connections between power and control.[34] The feeling is also well expressed when philosophers refer to a "will-to-power" in human behavior.[35]

Arousal related to experience of powerful machines is connected in some men's minds with sexual arousal, because with a woman, as with a car, such men like a power relationship in which they are in control. Advertisers acknowledge this by the kind of female figure they introduce into pictures that promote powerful vehicles. Even before advertisers had developed this theme very far, though, E. M. Forster used automobiles in his novel *Howard's End* as a means of portraying men who enjoy "the exercise of power without concern for the consequences," especially when driving with women passengers.[36]

A television sketch once dramatized the issue by showing a psychoanalyst trying to unscramble a car driver's "addiction" to (or dependence on) "the most environmentally unfriendly thing we do every day."[37] The driver was encouraged to talk about the way his "virility" as well as his enjoyment of speed and engine power were involved. Not only motor vehicles, but also furnaces, aircraft, and guns have been known to give people this kind of thrill. Designers of nuclear weapons and the X-ray laser "get high" not only when faced with puzzle-solving challenges, but also because of their "sense of deciding the destiny of the planet." Debra Rosenthal found some who were excited by the enormous power of a nuclear explosion. One expressed regret that the testing of nuclear weapons was at that time always underground. He wanted to "feel the heat" of the explosion—to see the flash, "brighter than a thousand suns"—and enjoy the experience of a massive energy transformation under human control.[38]

The power motive in technology has always been obvious, but the point here is that it should be considered alongside the puzzle-solving and frontier motives as part of what makes problems technically sweet, and what leads people to work compulsively on certain kinds of projects. However, scientists in weapons laboratories who seem to be influenced by the power motive also tend to have characteristic political views, exulting in the military as well as the physical power that nuclear technology gives their country.[39] By contrast, scientists challenged more by the intricacies of puzzle solving than by the exercise of power are less likely to link their compulsions to political purposes, and in some instances may seek to avoid work on weapons.

The Faustian Ethic and Compartmentalized Thinking

The drive to be always inventing, or applying whatever is on the threshold of possibility, often referred to as the technological imperative, is one of a cluster of values that has been described as a "Faustian ethic."[40] Other values that belong here include the power motive just described, and willingness to take risks for the sake of knowledge or power in what is often spoken of as a "Faustian bargain."

These allusions refer to a real if shadowy figure who practiced alchemy in Germany prior to about 1540, when he was allegedly killed by one of his own experiments. By the time an account of his life was printed, some fifty years later, many legends had accumulated around his name, some of them originating from events associated with other alchemists and "magicians" (such as Roger Bacon and Agrippa von Nettesheim). But central to all the stories was that Faust (or Faustus) sold his soul to the Devil in exchange for magical power with a technological flavor, the latter evident when he says:

Yea, stranger engines for the brunt of war . . .
I'll make my servile spirits to invent.[41]

The story has been used by several modern writers on technology to illustrate the power motive. For one of them, nineteenth-century engineer I. K. Brunel seems very like Faust in his grandiose schemes for engines, bridges, and ships.[42] For another, Edward Teller as the nuclear scientist always in pursuit of the ultimate weapon was following in the footsteps of Faust.[43] By contrast, engineers and scientists often find more comfort in the new ending that Goethe gave to his version of the Faust story in the second part of his drama, written many years after the first. In it, Faust discovers that there is no real satisfaction in the exercise of magical power, and devotes himself instead to a humble but socially useful engineering project.[44]

The Faust myth illustrates several common attitudes to technology. One is the attitude seen on a fairly trivial level, when a man who drives fast in a powerful car is happy to make an exchange of slightly greater risk for enjoyment of his engine's power. Those who promote nuclear power or develop more powerful weapons display the same attitude. In many instances, there is an arguable case that some of these Faustian bargains are worth making: for example, when it is judged that the enhanced electricity supply from a nuclear plant is worth the risk involved.[45]

The Faust myth illustrates another cast of mind evident in the practice of technology: the tendency of some individuals to become so absorbed in technical work, and go at it so compulsively, that for much of the time their attitudes become locked into one mental compartment. Then, like the Los Alamos scientists mentioned earlier, researchers may forget what empathy is, and lose touch with people-centered thinking.[46]

David Noble made the same point about compulsive problem solving work, as we saw earlier. For a person to be deeply absorbed in tackling a puzzle or exploring frontiers of knowledge can be "the well spring of creativity." But when such absorption "is indulged beyond reason, in defiance not only of personal health but of the larger social welfare as well, it becomes madness."[47] In other words, there is a Faustian bargain to be struck on a personal level as well as in terms of physical risks and dangers. One pays for creativity by risking one's own mental well-being.

Psychologists might question Noble's choice of words here, but some do say that mental compartmentalization arising from specialized work can be pushed so far as to verge on the "schizoid."[48] Alternatively, it can be represented as a "doubling" of personality in which one's technological self begins to lose touch with one's more people-centered, everyday self.[49] This is another way of talking about the situation described in previous chapters in which creative people get stuck in an object-centered mode of thought and the ability to switch across to more people-centered concerns is diminished.

Robert Jay Lifton uses the concept of doubling in his account of doctors who worked in Nazi death camps during World War II, because it can help explain how they so fatally lost touch with all vestiges of empathy or other human feeling. Although many factors entered into the doubling of personality the doctors exhibited, especially ideology, a major factor was their willingness to focus on technical and bureaucratic detail, coupled with a pretense of doing scientific work using the advanced equipment provided for research. Some of the doctors sought to remain in a narrowly technical frame of mind by working very hard at the minor practical tasks that fell to them, "simply absorbing oneself in medical work" for fourteen or sixteen hours each day.[50] But some men who occupied the medical/technical half of their minds very fully, quite clearly hated the work they were doing, and morally condemned it in some other mental compartment, resulting in markedly contradictory behavior. One doctor saved many lives through conscientious medical work and showed kindness to individual prisoners, but then contributed to the deaths of some of those he had saved. Others worked hard simultaneously both to keep the system going and to undermine it.[51]

Lifton makes an explicit comparison between these instances and the doubling of Faust's personality following his pact with the Devil. In Goethe's play, Faust is made to say:

Two souls, alas! are lodg'd within my breast
Which struggle there for undivided reign.[52]

One might equally cite Christopher Marlowe's play on the same theme, which shows Doctor Faustus responding first to a good angel and then to another, bad angel. The moral is that an individual who gives himself (or herself) up to compulsive, narrowly technical or bureaucratic work and closes off other compartments of the mind is in fact making the same sort of bargain that Faust made.

But it is not necessary to be so narrowly focused and compartmentalized to be creative as a scientist or technologist. Among the scientists who contributed to the development of the first nuclear weapons were certainly some who seem to have given themselves up entirely to object-centered work (probably Edward Teller); some who behaved in an ambivalent, if not contradictory way, as if they had "doubled" (Oppenheimer); and a few who consistently expressed deep social concerns and disowned most of what was subsequently done with this new technology (Rotblat, Einstein), retaining some sort of people-centered perspective throughout.[53]

A third and more trivial way in which the Faust legend is relevant here has little to do with the personality of Faust himself, but relates simply to the aura surrounding his activities as an alchemist and magician. For despite the significance of later forms of alchemy for the development of science (Chapter 3), its earlier, more extravagant claims, suggestive of dangerous, nature-defying activity, are what people remember. Much of what is said today in a speculative tone about future developments in technology is in some ways similar, of course, but in early twentieth-century discoveries made in the new field of nuclear physics, there was an even closer analogy.

Two pioneers in this field who worked together, first in Canada and then in Britain, were Ernest Rutherford and Frederick Soddy. After their first discoveries about splitting the atom, they were struck by the thought that they had achieved precisely what alchemists had once claimed when they spoke of transmutation of metals. In June 1919, Rutherford published work showing that by bombarding atoms of nitrogen with alpha particles, he had transformed them into atoms of oxygen and hydrogen. Even before that, Soddy was commenting that this "modern alchemy" was less concerned with transmuting other materials into gold (or oxygen and hydrogen) than with converting their matter into energy. And Rutherford explicitly characterized his branch of physics as the "new alchemy." Both men warned of the Faustian bargain that nuclear research entailed, with Soddy later withdrawing from active involvement in the research because of his concerns. Instead, he applied himself to studies in economics, because he thought it more important to discover "the underlying causes of human folly than to make further progress in physics."[54]

Soddy was moved to this train of thought by experience of the First World War, which had shown him how double-edged science can be. The Haber ammonia process solved a food production problem for Germany through its use in fertilizer manufacture, but was simultaneously employed to make explosives. "Why were the discoveries which science made for the benefit of mankind in every case used for its destruction?" Soddy asked, fearing that his discoveries about the atomic nucleus would be similarly misused. Freeman Dyson gave half the answer to Soddy's question when he asked another: "Why does war have to be so damnably attractive?"[55]

Dyson is a mathematician and nuclear scientist who has also written about Goethe's *Faust*[56] and must be well aware that his question and Soddy's refer again to our tendency to make Faustian bargains. Such bargains are entailed, indeed, in almost every major technological project.

In civilian life, for example, a different aspect of the alchemy of power finds expression in massive construction schemes. Even where nuclear technology is concerned, Alvin Weinberg once famously compared modern reactors and particle accelerators with the pyramids of ancient Egypt and the cathedrals of medieval Europe.[57]

A more recent writer on "macro-engineering" has argued in greater detail that every civilization can best be judged by its success at planning, implementing, and managing very large technological undertakings. But rather than dwelling on the achievements of the past, this author looked forward to the next very large projects that he thinks ought to be attempted—maybe a Star Wars defense system or an undersea tunnel linking continents.[58]

But there is more to large-scale projects than even those authors acknowledge, as David Nye showed in interpreting nineteenth-century achievements such as the building of skyscrapers, long-span bridges, and America's transcontinental railroads. These constructions evoked awe and a sense of the sublime in celebrating human mastery and control.[59] By contrast, Samuel Florman, the civil engineer, referred to the second part of Goethe's play where Faust turns away from the grandiose and powerful to seek satisfactions in humbler, more socially useful work. Florman claimed that the deepest satisfactions of the civil engineer come from a "sense of helping" and of contributing to the well-being of his fellow humans.[60]

Similarly, Martin and Schinzinger argued that ethical obligations concerning the contribution technology makes to society represent "positive ideals" that come naturally to engineers. However, such statements rarely take account of the strength of compulsive drives that come with puzzle-solving or frontier challenges, or with the effort to control powerful forces. They tend, then, to omit understanding of how easily the engineer's wish to contribute to human well-being can then be left behind. In one brief passage, though, Martin and Schinzinger broke through conventional complacencies about the engineer's positive social ideals by quoting Florman's comment that sometimes a project "bewitches the engineer." There are also "mammoth undertakings" whose appeal to "human passion . . . appears to be inextinguishable."[61] This talk of engineers being "bewitched" is as clear an admission as one will find of the Faustian dilemma in which technologists may become entrapped.

Perverted Technology and Enjoyment of Violence

This discussion has so far made it appear that violent, destructive, and lethal applications of science and technology are undesired goals toward which we may be drawn unaware by becoming enmeshed in bureaucracy, or as a result of distancing, or while we are distracted by absorbing and demanding work. The last, indeed, may come to have such a hold on us that we work in a compulsive way, not thinking of consequences. The allure of technological adventure, risk taking, and powerful machines may also lead to violent consequences, but these again are usually unintended. The compulsive use of power does not by itself explain the most violent applications of technology, some of which imply more twisted motivations.

For example, there is a form of quasi-sexual arousal, catered for by various kinds of pornography, in which violence leading to injury or killing is a positive stimulus. In 1997, there was much discussion of a film entitled *Crash* (based on a novel by J. G. Ballard) that portrayed car crashes leading to arousal. In the early years of the twentieth century, a group of mainly Italian and French artists and writers who called themselves Futurists explicitly linked such pleasure in violence to technology. They used violent images involving machines, electrical equipment, and weapons, and in a manifesto issued in 1909, proclaimed without hesitation: "We wish to glorify War." Painters associated with the movement were more interested in capturing the effects of electric lighting than in mechanical forms, and said that they felt "empathy with the World of Things." In 1910, five painters issued another manifesto that mentioned their "whirling life of steel, of pride and of speed." They expressed "contempt for women" as they sang of adventure, risk taking, "the love of danger, the habit of energy and boldness." With the outbreak of war in 1914, most of the Futurists joined the armies of their respective countries, and many did not survive.[62]

Another, mainly literary group emerged in Germany after the war, centered on a body of former soldiers known as the *Freikorps* who fought revolutionaries and workers during the civil disturbances of 1919 to 1923. Several wrote novels expressing militaristic fantasies, and some later became Nazis, including one who rose to be commandant of Auschwitz.[63] Analyzing their writings, Klaus Theweleit noted that their pleasure in violence was "massively intensified by the war's machinery." He quoted one writer who found the peacetime machine a disappointment because it delivered only "meager quantities of the intense pleasure of domination."[64]

For the *Freikorps* writers, as for some Futurist artists, attitudes to women were reminiscent of those of men who, three centuries earlier, had hunted alleged witches and had them executed. For such men, it seems, "woman" was the code word for a complex of feelings they feared, including feelings about nature and about the intuitive and the intimate. In Germany during the 1920s and 1930s, the *Freikorps* men suspected most women of being communists—Rosa Luxemburg was a particular hate figure—or else of being whores. In either case it was permissible to enjoy killing them. The few women who were tolerated were those who came in neutral guise as either sisters or nurses. Some of the *Freikorps* married the sisters of soldier friends; others turned to their machine guns instead, one describing feelings of near orgasm as "the gun wriggled and jerked like a flash. . . . I held it firmly, tenderly . . . clamping its tossing belly firmly between my knees." Klaus Theweliet added that these men were so violent that, when one of them described it as a "pleasure" to think of a bullet penetrating "warm, living human bodies," that need not be a sexual fantasy so much as enjoyment of blood and violence for its own sake. Such men do not need distancing to be able to commit acts of violence.[65]

The American edition of Theweleit's terrifying book on this subject is interspersed with illustrations culled from later comic strip portrayals of violence published in the United States, and from other pornography. In her foreword to this edition, Barbara Ehrenreich asked whether the attitudes of the German *Freikorps* were peculiar to a warrior class with experience of the First World War, or whether they have equivalents in modern societies, and among ordinary "normal" men who betray sadistic attitudes to women in their jokes.[66]

Some answers to this question were offered by Joan Smith, who found close similarities between the work of *Freikorps* writers and songs written (c.1980) by USAF pilots flying nuclear-armed F1-11 fighter-bombers. Again, technology provides metaphors of orgasm, this time in the form of the "flaming metal" of the F1-11 and the mushroom cloud of a nuclear explosion. Again, women are referred to with loathing as whores, apparently through fear of the strong, unmanly emotions they evoke. Joan Smith also pointed out films that dwell on rape and murder of women as if these crimes were punishment for whorelike behavior. Practical consequences have sometimes followed, she noted, when police with similar attitudes fail to understand crimes against women and do not take appropriate action.[67]

Others have seen a Futurist-style obsession with technology as a source of eroticism, violence, and death in the Strategic Defense Initiative, or Star Wars project, launched by President Reagan in 1983. William Broad's study of scientists working on the project certainly documented a masculine ethos in their laboratory and a preoccupation with death.[68] Edward Teller, the most prominent advocate of Star Wars, has already been cited for the wide range of spectacularly violent technologies he promoted. He first became prominent because of his leading role in developing the thermonuclear "superbomb" (or hydrogen bomb), which was tested in 1952. Not only did his later work concern other nuclear weapons such as the neutron bomb, but he also promoted "peaceful" (although violent) uses of nuclear technologies, notably for excavating harbors and reservoirs, and most recently, in 1996, for firing nuclear bombs into space to divert asteroids that may threaten to strike the earth.[69]

How Is This Possible?

The questions that finally arise are, How are such dangerously violent enthusiasms possible for civilized human beings? What in our society allows individuals even to think of such things as have been described in this chapter? And is there a systematic link with science and technology?

At one time it was often said that most animal species do not kill members of their own kind, and that humans are unique in their murderous tendencies. However, it is now realized that certain kinds of animals, including lions and hyenas, do sometimes kill other individuals belonging to the species. Moveover, this can happen with particular violence among chimpanzees, a species close to humankind in evolutionary background and genetic endowment. The chimpanzee, indeed, "stands at the very threshold of human achievement in destruction, cruelty, and planned intergroup conflict."[70]

When humans seem to enjoy aggression or violence, it can sometimes seem like play that has gotten out of hand, and it is sometimes suggested that this is a holdover from childhood fighting or from juvenile enjoyment of risky or violent adventure. Jane Goodall has discussed this in comparing young children and immature chimpanzees. There seem to be close parallels in the fighting behavior of these two groups when they are very young. But as children grow older and master language, they also learn the art of negotiating, and then their development diverges markedly from the chimpanzee pattern. They learn that human culture offers many other options for relieving tensions and settling disputes, and they may also learn how to be assertive without being aggressive. The predisposition to violent behavior remains, but in ordinary civil society, it need not be developed.

Studies of chimpanzee aggression and related work have led to the emergence of a distinct biological perspective that differs markedly from the view of aggression that psychoanalysis has tended to foster. Thus, it used to be thought that aggressive impulses and intentions are always in the background—in the subconscious—and need to find nonlethal outlets in competitive sport, or through drama involving some violence. However, many biologists now argue that we are endowed by nature with only a *predisposition* to aggression. The more extreme forms of aggressive behavior and violence have to be learned, and passed on within a culture. Some biologists also point out that there is another ape as closely related to humans as the chimpanzee, but lacking the aggressive tendencies of the latter. This is the bonobo or "pygmy chimp," and the point is made that among bonobos, the females are better organized and have more power within the group than among chimpanzees. Also, sexual behavior (including much homosexual activity) provides means of relieving tension without aggression.[71]

Bonobo behavior therefore challenges assumptions based on the usual "macho evolutionary models," and suggests a form of development that may be possible for humans, but that we have hardly begun to explore. Even if this were wishful thinking, however, there is a good deal in research on animal behavior to show that genes do not dictate that human males follow chimpanzee standards of behavior, even if these standards have been the dominant pattern among human males in the twentieth century.

Such thoughts, however tentative, should encourage us to ask what in our culture—including our political culture—leads us to turn a predisposition to aggression into episodes of extreme and cruel destructiveness in so many instances. Some explanations may be found in military institutions,[72] or in the way various fundamentalisms, ethnic, political, and religious, seem to make it feel right to exclude from society or even kill other people, just for being different. However, we ought to revisit briefly three other partial explanations, already mentioned in connection with attitudes to technology.

The first of these is compartmentalized thinking, which Robert Romanyshyn illustrated by tracing the history of the dissection of human bodies for medical research. He noted that this kind of scientific investigation at first depended on an ability to seal off all traditional feelings of horror in the face of death, and to separate knowledge from feeling by placing them in separate mental compartments.[73] A similar detachment from normal feeling is essential in certain jobs, such as butchering and slaughterhouse work, and is deliberately taught to infantry soldiers.

Infantrymen have to be able to kill in face-to-face combat, for example by a sudden knife thrust upward, under the rib cage. Part of the barbarity of the twentieth century is that sometimes children have been given such training. In 1995, children were found to be fighting in some thirty-three wars around the world, and some had also been employed as assassins in terrorist actions. They were said to be highly effective as battle troops because they were less restrained by fear or moral convention than adults. Moreover, one technological "advance" that has made this possible is that assault rifles are now so light that children of ten have no difficulty with them. The American M-16 rifle has been described as "the transistor radio of modern warfare" because it is so portable and available.[74]

A second issue is illustrated by an observation of anthropologist Paul Richards that in parts of West Africa, violent American films seen on portable video equipment are used to help prepare young people for a role in warfare. Adults in Sierra Leone have said that having watched the films meant that they were less surprised by the violence that engulfed their area in April 1991, and even learned some "tricks you need to survive." However, rebel leaders "seized upon . . . youthful enthusiasm for violence-as-entertainment as a recruiting ploy," and set about converting it to "violence-for-real by embroiling young enthusiasts in terrorist atrocities."[75] This is just one extreme example of a general question about how cultures of violence develop, and what role films, pornography, toy guns (and other war toys), video games, and television play. There will presumably always be controversy about this, and it is probably wrong to think of violent entertainment as an immediate cause of violence. But it is hard to disagree with Joan Smith's point about violent films and television appearing to give permission for one to enjoy, if not practice, certain kinds of aggression. It is also significant that a good deal of fictional violence makes a connection between powerful technologies and aggression against women. It appears that a certain kind of male fantasy links the two kinds of violence, technological and misogynist.[76]

A third point is suggested by one interpretation of the history of science, technology, and medicine, that emphasizes how these areas of knowledge and skill were consistently developed as masculine pursuits, especially during (and since) the scientific revolution. Writers who have traced this attitude from around 1600 to the modern period include

Carolyn Merchant, Robert Romanyshyn, and Brian Easlea.[77] They see in general two main reasons for the masculine emphasis. First, science and technology appeal to individuals who prefer object-centered work because they fear emotion, sexual desire, and the loss of control to which these passions lead. Another significant point about such people's lives is that they often have little contact with the area of life that has to do with children. Romanyshyn interpreted this, and the object-centered attitude, in terms of a "denial of the feminine . . . now regarded as alien and other." Some commentators have identified the problem as the male's awareness that he is not the full "equal of the female," because he cannot conceive and "bring forth life."[78] So when a man creates something practical, it may be described as "his baby," or he may be said to have "fathered" it, even as Edward Teller was "father" of the hydrogen bomb.

More simply, it has often seemed in this chapter that it is "fear of unmanly emotion" that makes men want to act tough and behave violently. The same fear, I suspect, makes academic commentators reluctant to recognize that personal feeling should be taken into account in understanding how individuals contribute to the creation and use of technology. When I reflect on my own experience of study and research in science and engineering, I have to say that there was little on the surface expressive of interest in domination, control, or aggression. Instead, people-related concerns were simply left out of account, and there was a mind-numbing lack of scope for individuality, imagination, or passion. And my sense of the passion that should have been there has been the driving force behind this book.

One reason for this chapter, then, is a conviction that mind-numbing philosophies of technology, and the attempt to deny passion, are not a source of objectivity but of distortion. Such philosophies, moreover, to some extent reflect a gendered outlook that the attentive reader will have noticed being discussed throughout this book. In chapter 2, for example, a series of women psychologists were quoted because of their concern over the object-centered bias of so many male scientists.[79] Robert Romanyshyn spoke of the need for a "reappearance of feminine consciousness in . . . our technical culture."[80] By contrast, Liam Hudson and Bernadine Jacot[81] have implied that this might be a false hope because, they think, the sexual division of labor is innate, and science and technology are a natural venue for the male imagination. One may question the word "innate" in this last statement, but accepting the general drift, we can see why philosophies of technology that ignore human passion and are blind to issues of gender do not make technology value-neutral, but bias it toward its negative potential.

Elsewhere, I have spoken of music as representing much of the passion and feeling that underlies technology, but George Steiner offered a sobering warning when he reminded us of Nazi men who "were avid connoisseurs, and in some instances, performers of Bach and Mozart," but who at the same time, devoted their working lives to organizing the delivery of Jewish and gypsy families to the gas chambers.[82] Similarly, William Broad documented pronounced musical interests among scientists working on the Star Wars project.[83] In such cases, it seems, the formal structures of music can be enjoyed, perhaps in a technical way, within a compartmentalized mind. But I have known occasions when music has had a powerful, mood-changing, and compartment-shattering effect, and has lifted me (at least) out of narrow preoccupations and object-centered moods. That has often happened when the music has been used in a social context, and perhaps especially where there is the mutuality of choral singing.

Such experiences give me some hope that despite the pessimism of George Steiner and the others quoted, something can be done to begin creating a culture in which violence (and its enjoyment) does not thrive, in which technically minded individuals are sensitized to people and grow in "moral imagination,"[84] in which men learn to value authentic human feeling rather than dismissing it as unmanly, and in which the male of the species begins to understand love as cherishing, as well as love as sex.

But the political context of such a culture has to be congenial to the reduction of violence as well. Manifestations of nationalism, of fundamentalism, and of illusions of ethnic or moral superiority are inimical to civilized human behavior, and so also is the current dogma about globalized free markets. The latter not only foster ruthless economic behavior but also have a distancing effect, so that those who make profits never see those who bear the costs. Without any overt use of violence, an attitude of disregard rather than the cherishing of people is encouraged.

John le Carré, novelist of the cold war and hence a keen observer of the old Soviet Union, has commented that perestroika of the kind initiated there by Mikhail Gorbachev did not even begin in the West. Yet he regards our free-market system as in as much need of reconstruction because, in many parts of the world, it is "a wrecking, terrible force, displacing people (and) ruining lifestyles, traditions, ecologies . . . with the same ruthlessness as communism."[85]

III

Conclusion: The Missed Opportunity?

9

People-Centered Technology

Process and Design

Modern technology seems to reflect a predominance of what previous chapters characterized as "object-centered" creativity among its practitioners. This raises the question, What would technology be like if, over a long time, it had consistently been developed by individuals whose outlook was "people-centered," and whose idea of what life is about was centered on love and care for others?

There are many directions in which we might look for answers to such questions. Ergonomics quite deliberately focuses on human body shapes and movements in order to design machines that are convenient and safe to operate, or to produce chairs that are comfortable to sit in. This is certainly people-centered work in a limited way, but it is not much concerned with human relationships or personal values.

We might also see work on the safety of food and drugs as an important aspect of people-centered technology, and we might regard the designers of such things as seat belts for automobiles in this light, admiring especially Ralph Nader's campaigns. However, remedial work aiming to repair the inadequacies that arise in object-centered technology is not a clear guide to what technology could have been like had there been a focus on people from the start.

What may take us closer to the heart of the matter is practical work carried out in situations where cherishing others is an unambiguous purpose, even when this includes such inconspicuous activity as cooking for a family, to which the term "technology" applies only marginally—

unless, perhaps we redefine technology in such a way that an art such as cooking becomes a central example of how technique and skill ought to be related to human needs. Such a redefinition may, indeed, be desirable if we are to understand the implications of a people-centered approach.

One further example to consider, though, is the work of innovators who have deliberately attempted to put people at the center of their endeavors and who describe their work as "appropriate technology" (AT).[1] One oddity about this sphere of work, and also about cooking, is a tendency to refer to technology as "economics." Thus cooking, with other domestic technology, is sometimes described as "home economics," and Schumacher launched his version of appropriate technology with "a study of economics as if people mattered."[2] The use of that word may reflect an unconscious reaction against the focus on hardware encountered in object-centered discussions, and a wish to think more in terms of the processes of providing for a family or serving the needs of a community.

The essence of a people-centered approach, however, must lie in the relationship between the technologist and the people who use or benefit from the processes or techniques he or she develops. An example is provided by an AT project in a drought-prone region of Africa where an agricultural engineer was sent to initiate improvements in soil and water conservation. His aim was to make it possible for crops to be grown, despite near-desert conditions, and he had experience of Israeli water conservation techniques that seemed likely to be relevant. But he also took the trouble to learn the local language, and in that way formed genuine friendships, and began to understand the local way of life. In this respect his outlook was strongly expressive of a Baconian "love of others," and that made him aware of local reactions to the techniques he was trying to introduce and awakened him to processes of water conservation already being practiced. Slowly, through discussion and experiment, new ways of using rainwater to grow crops were evolved, quite different from the Israeli model and of a kind related to techniques that people already knew. So modest innovation in low-technology conservation methods grew out of the friendships and dialogue between the community and the engineer.[3]

Some such dialogue, based on people's reactions to one another's technical experience, is an essential part of most people-centered technology, as other commentators show by reference to traditional relationships of craft workers and the communities they served. Eugene Ferguson pictured a local boatbuilder in regular contact with the fishermen who bought his boats, and envisaged one of them saying to him: "Joe, that damn boat almost killed me crossing the bar today. You've got to cut away the forefoot a little and put some beam into her, back aft." Ferguson commented that whether or not the builder found it possible to modify that particular boat, the next one he built would be different. The local type of fishing boat would thus evolve over time by an extended dialogue between users and builders.[4]

The same kind of process went on among the English wheelwrights often quoted as an example of craft design (Chapter 3). They were said to be so much a part of the local community that they knew each customer (and his horses) well enough to modify standard wagon designs, even before the customer asked, to suit his farm or business. This illustrates again what dialogue can mean, and can be compared with the communal, ritualistic way in which North American Indians traditionally made canoes together, and the Inuit made kayaks.

Such examples of close contact between makers and users also fit Ivan Illich's notions of there being a "convivial" way of practicing technology. Although Illich intended that term to have a specific meaning designating a "modern society of responsibly limited tools," he also wished the concept to denote an involvement of ordinary people in the practice of technology as the foregoing examples illustrate. Implied as well is the enjoyment that can come when people work together.[5]

Many related experiences are discussed in the context of modern industrial design by such writers as Christopher Jones and Victor Papanek. Jones is notable also for his references to traditional technologies such as the work of wheelwrights making carts and wagons. Commenting on designs of wagon wheels, he wrote of the "beautifully organized complexity" of traditional wagons, which he compared with the subtlety of design seen also in small boats of traditional construction, or in handmade violins. He suggested that such craft products are like "naturally-evolved forms," their design having developed over a very long time by processes of gradual adjustment to the needs of users, and to the qualities of the materials used. They are also elegant in the mathematical sense, even though resulting from communal construction.[6]

The problem for the modern designer is to achieve as good a result in a much shorter time and in a situation where dialogue between users and makers of products is no longer an everyday occurrence. Thus attempts are made to collect consumer reactions through surveys, response cards, focus groups, and the like. But even where these mechanisms work, they cannot work in the same way. Traditional design concerned itself with local people and responded to local needs. Now global comment is invited, and designers attempt to merge local preferences. Victor Papanek [7] discussed design in this context but underlined his people-centered approach by emphasizing that individuals and groups have specific needs that cannot be merged, varying with gender, income, home circumstances, and ability (or disability). This is comparable to the way Schumacher and other advocates of appropriate technology have been able to illustrate people-centered approaches in technology by discussing particular needs in less-developed or third world countries. Simply looking at a product or process in a different cultural context, or in relation to untypical circumstances in one's own community, can help increase awareness of how technologists should respond to people.

Indeed, there has been some cross-fertilization between Schumacher's approach and Papanek's, although the latter's starting point was the design of consumer products, while Schumacher focused more closely on work and employment. Work, Schumacher said, has three objectives: to produce necessary and useful goods and services; to enable us to use and develop our abilities, skills, and other qualities as people; and to provide the means for individuals to collaborate and cooperate with one another.[8] If only the first of these objectives mattered, then it might make sense to use machines to produce all the goods we need with the lowest possible number of workers, but a rounded view of human development calls for attention to the other two points as well.

Victor Papanek added the further comment that market forces take little account of some human needs, especially when people do not have the resources to express their needs as a demand in the marketplace. He saw market-driven design and technology as consistently failing to meet some needs, therefore, and suggested that designers might wish to give a proportion of their ideas and time on an unpaid basis for use in a people-centered mode. It is interesting that Martin and Schinzinger, in discussing the ethics of technology, see a role for engineers also to take on voluntary work of this kind.[9]

Industrial and Medical Contexts

The examples of people-centered technology mentioned so far refer mainly to small-scale or traditional technology. When we turn to large-scale industry and high technology, what is striking first of all is the considerable volume of debate about how "soft issues" connected with job satisfaction and employee welfare should be related to "hard issues" in engineering, system design, and scientific method.

In many activities dependent on technical expertise—medicine or water supply as well as industry—taking account of the human aspect seems always to present problems. One writer on information technology defined a "technological domain" comprising the computer system that processes data and a "people domain" consisting of the individuals who interact with the data to interpret it and so produce information.[10] There are clearly several incompatibilities between these two domains that may be exacerbated if systems designers apply an object-centered approach (as defined in Chapter 2) throughout their work. The contrast between object-centered and people-centered ways of thinking should not be exaggerated, but it remains true that many jobs cannot be done adequately by focusing entirely on objects—or exclusively on people. One has to be able to switch from one to the other.

For example, a nurse cannot take a simple object-centered view of the human body if she (or he) is to feel empathy with her patients and care for them well. Yet in a modern hospital, nurses may spend a great deal of time using computers or working as technicians with complex equipment. An article in *Technology Review* describing a hospital in Massachusetts set out to emphasize this aspect of nursing, but also attempted to demarcate the boundary between technological and caring skills. It described a patient who is unconscious following a surgical operation, no doubt because at that stage, nursing can indeed be almost wholly object-centered. But as the patient wakes up, the article noted, the nurse switches to "another critical aspect" of her job, namely, "caring for the human being who lodges in the body she has tended. . . ." The assumption of a mind-body dualism, with the body seen in an object-centered perspective, is very striking. But although an object-centered view of tissues and organs may sometimes be necessary, nurses should be able to see beyond the physiological horizon, as the article later said, to "the human experience of loss or dysfunction."[11]

But there are sometimes worrying signs that these kinds of more human focus are being eroded by a tendency in hospitals, as everywhere else, to employ machines in place of people. When fetal heart monitors are used in a maternity ward, some of the skilled work of the midwives is displaced. The result is that they observe the health of expectant mothers less completely, and get to know them individually hardly at all.

Psychiatrist Robert Jay Lifton put the problem of mind-body dualism into perspective when he said that there is inevitably a place for a "mechanical model of the body" in the practice of medicine. But the relationship between doctor and patient should be based on an understanding that the doctor has implicitly asked if he or she may "look at your body as a machine" for the time being, in order to do whatever is possible "for your overall health" as a person.[12]

Problems also arise for a people-centered approach when laboratories become ivory towers in which experts pursue their scientific interests without recognizing that they have lost touch with their proper people domain. Some then "labor in research facilities remote from clinics and nursing homes." To counteract this, one expert on Huntington's disease regularly brought patients and their families to academic conferences on the subject to meet the scientists. The hope was "that the researchers would ever keep in mind the very real suffering" their work was meant to address.[13]

There are several other barriers between experts and the intended users or beneficiaries of technology, one of which is due to the high status of "hard" technical disciplines as compared with "soft" human and social sciences. In applications of information technology (IT), for example, people who work on its psychological and organizational aspects often say that they "feel marginalized," despite evidence that the disappointing performance of many IT systems is due to neglect of the human dimension. One reason may be that the development of IT is driven by engineers, computer experts, and accountants, who are not always responsive to ideas from disciplines such as applied psychology. Within some engineering and accountancy circles, indeed, the usual view is that "new technologies provide an opportunity to reduce both staffing levels and skill levels" in industry and business. When performance is disappointing, the users of equipment are regarded as "sources of error and unpredictability." Then engineers respond to complaints from those who have to use the system by saying that problems can be eliminated if operators and users are, as far as possible, "designed out" of the system. Meanwhile, the operators themselves are usually very resourceful in coping with ill-designed systems, though of course, they tend to blame any failings on "computer error."[14]

There has, of course, been a trend to design more user-friendly products and packages for IT. However, according to Chris Clegg, from whom many of the above comments are quoted, even those arguing for people-centered design do not always carry through that approach consistently. Some are seduced by the "top-down" approach so tempting for any expert, who then gives to users "what we think they need." Beyond that, Clegg indicates that problems arise from opposed concepts of what technology is about. One concept is what he calls "user-centered"; another is based on the viewpoint of those engineers who aim to design out operators. This latter goal corresponds with the ideal mentioned earlier of making a system so perfectly automatic that it functions as a "world without people."

Alternative Paradigms

In comparing attitudes focused on designing people out, with those supportive of a people-centered approach in technology, we are comparing two distinct paradigms. Though discussion of paradigms in science has long interested philosophers and historians of the subject, recognition of different paradigms for technology is fairly recent.[15]

In this context, a paradigm is "a coherent and mutually supporting pattern of concepts, values, methods and behavior" that shapes the way a person looks at the world. Reviewing modern life with that definition in mind, Robert Chambers noted that the final decades of the twentieth century have seen parallel changes in many paradigms. In particular, there has been a move away from the rigid views that prevailed for a long time around the middle of the century, among Marxists and capitalists alike, and in social science as well as in business management. In all these areas, Chambers argued, more flexible paradigms have been developing that give more scope for dealing with the uncertainties of real life. He also noted parallel developments in science, citing chaos theory in passing, and then commenting on a number of environmental problems. "Global environmental issues involve huge uncertainties," he noted.[16] Such issues demand what Funtowicz and Ravetz called "second order science," in which judgment plays a more adequately recognized part, and there is a less simplistic reliance on calculation.[17]

The other contrast Chambers noted in his multidisciplinary review of prevailing paradigms is that the dominant ones tend to focus on things rather than people. That brings us back to the contrast, discussed intermittently here since it was introduced in Chapter 2, between object-centered and people-centered approaches to the practice of technology. It also brings us back to the work of Howard Rosenbrock, engineer and philosopher, who made the same kind of comparison with regard to how human abilities are valued. For example, during the 1970s, in writing about computer-aided design (CAD), Rosenbrock expressed concern that use of the computer to provide an automated design manual seemed "to represent . . . a loss of belief in human abilities," because designers were often reduced to making routine choices between fixed alternatives. To remedy this, Rosenbrock devised an interactive graphic display system that allowed designers to retain the initiative and develop their own skills.[18]

Similarly, in writing about a factory production line, as we saw in an earlier chapter, Rosenbrock commented that the abilities of the people who worked there were treated with less regard than the capabilities of robots. And of course, this same tendency to undervalue people's abilities is reflected in the attitude that unthinkingly blames every failure of a computer system on operator error and never on, say, design faults.

Long experience of engineers with this attitude led Rosenbrock to think that the problem lay with a paradigm for technology that is never made explicit, but nonetheless is picked up by most engineers during their training, along with other unquestioned values and assumptions. In some respects, this paradigm is another expression of the long-standing fascination with automata and worlds without people discussed in Chapter 7. But in one part of his analysis, Rosenbrock traced some aspects of this paradigm to a specific origin during the industrial revolution in Britain.[19] At that time, he argued, mechanical inventions were of two kinds. One was typified by Hargreaves's spinning jenny, a manually operated machine invented during the 1760s for use by people spinning cotton yarn in their own homes. It allowed several threads to be spun simultaneously, and it led to higher production than was possible with a spinning wheel, which could only spin one thread at a time. However, the machine could achieve this higher production only by requiring greater skill and concentration of the operator.

By contrast, other inventors of the period aimed to produce spinning and weaving machinery that would be used in the factory, not the home, and would require operators to deploy less skill—so little, indeed, that some of the machines could be worked by children. Some looms and spinning machines were described as "self-acting," and were to a high degree automatic. In 1835 a book on the "philosophy of manufactures" was published by Andrew Ure, who was something of a spokesman for the factory owners. This book proclaimed that "productive industry should be conducted by self-acting machines" and that the "most perfect manufacture is that which dispenses entirely with manual labor."[20]

In other words, the self-acting loom or spinning machine was seen quite explicitly as a step toward a world without people within the walls of a factory. In more practical terms, Rosenbrock noted, conditions prevailing in the nineteenth century meant that inventions of self-acting machinery were always more profitable than inventions on the Hargreaves model. In other words, machines that led to displacement or deskilling of labor paid off so well that they came to be seen as the goal of any inventor or engineer interested in productive machinery.

What is especially interesting about Rosenbrock's work, however, is that he presented the spinning jenny as illustrating a paradigm for skill-enhancing invention, something that he also demonstrated with his interactive computerized design system, and later, with an integrated manufacturing system that was set up in a factory making electrical connectors.[21]

A further point emerges when we note some extra detail Rosenbrock gave about Hargreaves and the spinning jenny.[22] He commented that Hargreaves first made the jenny for his own and his family's use (in 1764), at a time when cotton yarn was being spun at home by women (mainly) using spinning wheels. Hargreaves's invention enabled spinners to keep pace with increasing demand while still working in their homes. By 1768, he had made so many jennies that people felt their jobs threatened, and by this time, Hargreaves seems to have been thinking like an industrialist. But his initial concept was related to home-based production, and was focused on his own family.[23] So this was an invention originating from within the social context where it would be used, as compared with inventions that originate from an engineer's external view of a system, or from top-down ideas of what is good for other people.

In this respect, Rosenbrock's account of Hargreaves bears some resemblance to Patricia Thompson's interpretation of pioneer work on home economics in the United States around 1900, which again focuses on science (or invention) originating within the situation where it was to be used.[24] The founder of the home economics movement was Ellen Swallow Richards, who studied chemistry, then used her skills in work on foodstuffs and water supplies. For example, she analyzed numerous water samples for the Massachusetts water survey. And having been the first woman to graduate from MIT, in 1873, her achievements in this field resulted in her becoming the first woman to teach there. She had also done promising work in pure chemistry, but did not pursue that because of her concern about public health and her wish to put home economics on a sound basis.[25] Thus she made a deliberate choice to be involved in life, rather than becoming a laboratory scientist.

Among many other projects on which she worked, one was to advise Edward Atkinson about a simple, low-cost oven he had invented in which workmen could cook food at their place of employment. By 1893, some 600 of the ovens had been sold, and reading about it today, one cannot help feeling that this had very much the character of more recent appropriate technology inventions. Of particular interest, though, was not only the oven itself, but Atkinson's concern that it should lead to improvements in nutrition. Thus he combined instructions on use of the device with information about diet compiled "under the direction of Mrs. Ellen H. Richards."[26]

In the background of this work, however, were Ellen's writings on home economics and the domestic environment. Interpreting the whole corpus, Patricia Thompson argued that it represents an alternative paradigm, or conceptual framework, for science, because home economics is a science in which "the investigator and the phenomenon being studied form part of the same system." Whereas conventionally, scientists and engineers are assumed to be outside the systems they work on, discoveries in home economics, like inventions such as Hargreaves's jenny, come from within the system.

Patricia Thompson characterized this kind of involved science and technology as having a "Hestian paradigm." She derived this term for a home-centered science from the name of Hestia, goddess of the hearth in ancient Greece. However, scientists operating outside the home can also be regarded as using the same paradigm if their work is characterized by involvement in the system they are studying. Here Thompson cited Barbara McClintock, whose visual imagination took her "right down there" among the chromosomes she was studying so that she forgot herself entirely and felt that she was part of the chromosome system. At such times, McClintock's mood and outlook were characteristic of the participatory approach discussed in Chapter 3, in sharp contrast to the remote, detached way of looking at things that is supposed to be the scientific norm. Patricia Thompson made the same comparison by saying that when work is done in this spirit, learning, discovery, and invention arise from feedback within the system of which one is part. There are then no arbitrary boundaries between "science" and "life."[27]

One characteristic of technology practiced in a participatory way is that much interesting innovation tends not to be noticed. People who make inventions for use within their own circle or home environment are inevitably less conspicuous than the experts from outside. For example, Helen Appleton has documented recent work on traditional cereals grown in the Andes since Inca times. The aim of this work was to invent or discover ways of using the cereals with modern food processing techniques. Mixed flours have been produced suitable for making bread and cakes of high nutritional value and appealing to modern tastes. Had the work been done entirely by professionals in food technology it might have attracted considerable interest. But because it was done by a group of women making products for sale in local markets, the inventions and knowledge gained remain a largely "invisible technology."[28]

Patricia Thompson is more concerned with science than with such basic technologies, but even there, has suggested that it may be more common for women to adopt a participatory, Hestian approach. Apart from Ellen Swallow Richards and Barbara McClintock, she cited Rosalind Franklin (one of the discoverers of the double helix structure of DNA) and Rachel Carson (author of *Silent Spring*,[29] the seminal book on agrochemical pollution). However, she does not see this as "feminist science," and it is worth noting the parallels between Barbara McClintock and Michael Faraday. As we saw in chapter 3, Faraday also felt imaginatively drawn into the systems he studied, and hence practiced science in a distinctively involved manner. Indeed, one of his biographers commented that in his experiments, "Faraday was a participant," not just an observer, and his work gave him a feeling of intimacy with the physical world and with nature "herself."[30]

It should be noted that the approach to science and technology described here is participatory in two senses. Sometimes we are talking about participation of technologists with and among people. But sometimes we are speaking of participation of scientists in an experiment, or in the environment, or otherwise interacting with nature. In Chapter 3, the emphasis was on participatory relationships with nature, but here we mainly consider the involvement of technologists with the people they serve. However, designers, architects, and engineers often have to be participatory in both respects.

In architecture, for example, Christopher Alexander has argued that people should not only be consulted about buildings being planned for their use, but "should actually help design them." At the same time, he has suggested that design should be done at least partly in the actual environment where the building is to be erected. An example he cited is a mental health center in California that, about 1970, he was "designing on site with the client." One of the architects with whom Alexander should have been working, though, came to the site but soon said that he could not carry on like that. He could only "work at the drawing board," detached from both landscape and client. This architect was not "free enough in himself . . . to conceive the building right there and then," out on the site, or with potential users of the building.[31] Alexander and other colleagues aimed for a "new attitude to architecture"[32] that would allow houses and other buildings to grow out of human need, and from interaction with the environment—a strongly participatory concept.

Ecocentric and People-Centered Approaches

When we ask what technology would be like if it were practiced by individuals with a people-centered outlook, the main answer to have emerged so far is that it would be participatory in style, in complete contrast to the detached work done by those architects who only sit at drawing boards (or the electronic equivalent), or by many modern scientists. However, this question about people-centered (rather than object-centered) outlooks arose from other questions, near the beginning of the book, about individual experience, and about personal responses to technology.

A different kind of argument comes from commentators whose starting point is the environmental crisis of modern times, and who wish to describe how we ought to practice technology if we are to begin alleviating the massive problems represented by climate change, rain forest destruction, pollution of land and water, species extinction, and so on. Some serious thinkers on these subjects talk about "deep ecology" and call for an ecocentric (or biocentric) approach in technology.[33] They argue that humans are only one part of nature, and should adapt their ways of living to fit in with the other parts, rather than modifying the environment to suit themselves—rather than "remaking nature," indeed, or replacing it with the technical milieu discussed in Chapter 6. Deep ecology thinkers criticize the attitude of uninhibited willingness to modify large areas of landscape, to manipulate the genetic constituents of cells, and to engineer almost everything as "anthropocentric," because it seeks to further human dominion over the earth. It presupposes a paradigm of "humans as managers" in relation to nature.

Many advocates of deep ecology would probably regard the people-centered approach in technology advocated here as valuing humans too highly and nature not enough. They would tend to think that being people-centered also means being anthropocentric. However, if people-centered concerns are expressed through a participatory approach, that does not follow. The anthropocentric humans-as-managers paradigm can work only on the basis of a detached view of humans as separate from nature. It is not consistent with any of the participatory approaches discussed in this chapter.

An illustration might be the agricultural engineer working in a drought-prone region of Africa who was mentioned earlier. By responding to ideas from the local community, with its long experience of a semidesert homeland, he was able to fit his water conservation methods closely to the local landscape, building more modest earthworks than at first envisaged, and using local crop plants to produce remarkably good harvests.[34] The humans-as-managers paradigm, by contrast, would almost certainly have encouraged the view that much more radical modification of the environment was needed, because the system was still vulnerable to drought. That might have led to some kind of irrigation being introduced, despite the environmental obstacles and social disruption it entailed.

In practice, then, a participatory, people-centered approach to technology often converges with ecocentric ideals, even though the initial assumptions are different. But is the converse true? Can it also be said that individuals who start with a deep ecology stance converge on the same recommendations as individuals practicing people-centered technology? It has to be said that some deep ecologists and conservationists talk as if they don't like people very much, and would prefer almost everywhere to be a nature reserve from which humans are excluded. This is a worlds-without-people model in which the aim is a world safe for wildlife rather than a world populated by robots.

However, the most thoughtful writers on deep ecology do accept some people-centered values, if modified by ecological consciousness. They do, in fact, seem to appreciate that cherishing people and cherishing nature belong together. Freya Mathews, for example, has an essay on "value in nature and meaning in life" that approaches this point.[35] But the most compelling and moving statement of principle I have encountered comes from a biologist, Martha Crouch, and makes a clear link between ecological and people-centered concerns.

Crouch has argued that modern agro-industries, equipped with the latest biotechnology, do more ecological damage than any other major industrial sector, because they drive out "true cultural and biological diversity . . . in the name of development and efficiency." An example is the industry that produces palm oil, a major export commodity from tropical countries and one of the most important edible and soap-producing vegetable oils. At one time, many small farmers would plant and harvest the relevant species of palm, but today it is often grown on large plantations that represent a considerable capital investment. The capital required is even greater if genetically engineered plants are to be grown. "Poor farmers cannot . . . buy into this system, nor can they compete with its outcome." In just one country in South America, it was reported at the end of the 1980s that 115,000 people were threatened by oil palm plantations. For some it was a matter of losing their livelihood, whereas others were affected through the pollution of water by pesticides, and through the destruction of rain forest.[36]

Advocates of this kind of development argue that we need the new agricultural technologies if the world's growing population is to be fed. But my own reading and research on the green revolution in Asia[37] has confirmed Martha Crouch's conclusion that industry-driven innovation in tropical agriculture almost always impoverishes as many as it benefits. Food production may increase, but the numbers of people who cannot afford to eat enough increases also, and this is all achieved at heavy cost to the environment. Those who gain most from this sort of innovation are farmers and landowners who were already wealthy and shareholders in international companies. So this is not even comprehensively anthropocentric technology. Perhaps we should call it plutocentric technology!

That takes us into a political area, which is very necessary when it comes to trying to change things. But my initial reason for raising the issue concerns Martha Crouch's philosophical stance with regard to what she identifies as her ecocentric viewpoint. For her kind of ecocentric view turns out to be people-centered also, especially in the concern she shows about the lives of local populations affected by agricultural development. It is also relevant to note that, at the time she wrote her article, Crouch was pursuing a successful career in an American university. She was a researcher working on plant molecular biology. Moreover, her article expressing concern about the consequences of her research appeared in a specialist scientific journal,[38] where it was intended as a challenge to her colleagues to take a less detached view of their work. She also commented in more general terms about the gravity of the world environmental crisis, saying that she no longer felt that her own efforts to do something about it, "such as recycling or walking to work" were enough. A more appropriate response was "to change the kind of work I walk to." Thus she was quitting laboratory science and renouncing her chosen field of research to study instead how plant science can better serve humankind, and without ecological damage.

This is one way of reacting to what seems the great irony of modern science, that just at the point in history when it is revealing the full complexity and wonder of the processes of nature, it is at the same time serving interests engaged in the comprehensive destruction of ecosystems, species, and much that is needed for the continuance of life.

Not surprisingly, other biologists who quote Crouch's paper are not so alarmist. But some agree that there are major unanswered questions about "the motives, social responsibilities, conflicts of interest and accountability" of agricultural scientists. They acknowledge that because genetic engineering applied to crop plants "is a triumph of the reductionist approach to biological science," a danger may arise that similar "compartmentalization" will affect assessment of environmental consequences. To counter this tendency, scientific reports on innovation in plant genetic engineering could routinely include environmental impact assessments. Preparation of these latter would require "the participation in basic research . . . of personnel with a much wider range of expertise" than at present—personnel capable of challenging compartmentalized thinking, therefore.[39]

Martha Crouch, and thirty years before her, Rachel Carson, can be compared as biologists who have worked in research laboratories (and in Carson's case, in science journalism also[40]) who both reached a point where they needed to reject the detached, compartmentalized mentality of conventional science to take a stand with political implications. Others I know see potential benefits in technology that are not being passed on, and have left their drawing offices, workshops, or laboratories to work directly with the people who most stand to gain, and have thereby added a people-centered dimension to their work.[41]

Even without making such overt participatory links with people, technology and science can also be participatory in the sense defined in Chapter 3, and illustrated there by reference to several craft workers and also by such figures as Barbara McClintock and Michael Faraday. All these were individuals who felt personally involved with their materials and more generally with nature.

Each of these participatory ways of working—with materials, with people, or through political involvement—tends to be supportive of the others, and all are useful in their place as counters to excessive detachment.

Ethical Awareness

To talk about people-centered attitudes is to borrow a term that arose originally from studies by psychologists. They wished to describe factually how different individuals respond to science (and technology), and found quite distinct responses among many, who are described here as having "people-centered" interests. This chapter has also attempted to be factual (if rather anecdotal) in describing efforts to practice technology in a people-centered way. But however factual my descriptions, I have hardly disguised an ethical judgement that we ought to practice technology with a greater people-centered emphasis than is usually evident.

In this chapter, therefore, and throughout most of Part 2 of the book, I have been advocating an ethical view of how technology should be used, though without much direct comment on ethics or ethical principles. To conclude, then, it may be worth being explicit about some issues that were only implicit in earlier chapters, and considering what they indicate in general terms about meaning in technology. There are three main points to discuss:

First, many of the comments quoted from psychologists, in Chapter 2 and subsequently, are useful chiefly for identifying obstacles to ethical behavior in technology and in the practical application of science. Prominent among such obstacles is the habit of compartmentalized thinking that some people use to keep their object-centered interests strictly separate from people-centered sensitivities (Chapters 2 and 8). Often, compartmentalized attitudes seem harmless enough, except for the impoverishing effect they can have in the lives of individuals who become wholly absorbed in technically fascinating work (often these days in computing). Sometimes, too, a compartmentalized approach becomes institutionalized for "backroom boys" and the occupants of ivory-tower laboratories. The danger in these situations is that the detached way of working is a means of switching off conscience and disowning awareness of larger concerns. It is a way of *not* thinking about environmental destruction, military applications of technology, and under the Nazis, about what gas vans (for example) were used for.

Some psychologists would not be surprised that many of the exponents of people-centered technology cited in this chapter are women, because men so often prefer object-centered work.[42] Some men, though, begin their careers as object-centered thinkers, but then claw their way back, as they grow older, from youthful enthusiasms for things and systems toward a greater involvement with human affairs. Some return to a people-centered orientation only as they near retirement, and their reminiscences or memoirs then have an apologetic note. They explain how they were led astray in their youth, in one case by "technological exuberance" associated with work on the nuclear "superbomb" (the hydrogen bomb).[43] Clearly, we need a culture and a form of education capable of helping individuals of this kind rediscover their people-centered sensibilities more quickly.

Beyond the question of male propensities to object-centered thinking, however, there is also the problem of individuals who feel emotionally drawn to forms of technology that express aggression. Psychologists and biologists now seem largely agreed that "aggressive behavior, especially in its more dangerous form . . . is learned."[44] It is not an inevitable part of being biologically male, but may be part of the gendered culture of being masculine. Such culture propagates "a distorted idea of maleness" by celebrating military heroes, tough sportsmen, and "cowboys quick on the draw."[45] Some studies show in detail how pornography promotes a culture of male violence, often linking it to resentment of women.[46]

By contrast, efforts by psychologists to understand whether and to what extent men would be aggressive or violent in the absence of cultural influences show that most men are no more aggressive than most women. There are significant differences among a minority, of course, but more striking is the observation that men's aggression is less likely to be moderated by empathy for the victim.[47] Similar conclusions stem from recent experience of women in military service. Mixed-gender army units tend to be more effective in international peacekeeping work, because "female soldiers . . . display a compassion found less frequently among men," and may influence male colleagues toward more empathetic patterns of behavior.[48]

Empathy, compassion, and an understanding of what is meant by cherishing all help define the people-centered outlook with which this chapter has been concerned. Conversely, an object-centered viewpoint is characterized by inhibition in these areas. Accounting for the active role of many doctors in Nazi death camps, Robert Jay Lifton commented that medical training in Germany even before the Nazi era "produced powerful blocks to empathy and compassion," and added that this has been a problem at times in medical education almost everywhere. In a less acute way, it can be a problem for technical and scientific education also.

A second way of thinking about ethical responsibility in technology, though, is by identifying the paradigms or conceptual models around which values and unstated assumptions tend to cluster. Because technology (as distinct from pure science) is concerned with action, its paradigms always have ethical implications stemming from assumptions about how people and nature are to be valued.

Lewis Mumford once noted that, almost from the beginning of civilization, "two disparate technologies" (that is, two paradigms for technology) "have existed side by side." On the one hand, there was the technology of the pyramids and other great projects that reflect the power of autocrats, military establishments, or commercial empires. On the other hand, there was the technology practiced within families and local communities, which Mumford called democratic and others would call convivial—a technology in which Mumford detected the influence of women inventors and domesticators of food plants, as well as the work of men in many trades and crafts.[49]

Anybody reviewing more recent history with an awareness of the issues Mumford raised soon recognizes the influence of Francis Bacon and his use of the Genesis creation myth according to which "man" was to "have dominion . . . over all the earth," and was to "subdue it." This was one source of inspiration for several paradigms identified earlier: "remaking nature" (Chapter 6) and "humans as managers" (this chapter). But paradoxically, Bacon regularly criticized applications of knowledge and skill that were motivated by "lust for power"[50] and would have shared Mumford's dislike of pyramid-building projects. For him charity—as we saw in Chapter 7—was the central, people-centered purpose for which science, and what we now call technology, should be used. If science were to be "severed from charity, and not referred to the good of . . . mankind, it had rather a sounding and unworthy glory."[51]

In Bacon's work, then, we have the contradiction mentioned earlier between two paradigms for technology, a contradiction worth underlining because in many ways it is still with us. Thus we still need to be asking, Are our paradigms for technology sufficiently people-centered for words such as empathy, cherishing, or charity to be applicable? Or have ideals of detached analysis and study so taken hold that these participatory values are ruled out? Are our paradigms mechanistic or purposive (as these terms were used in Chapter 1), and how does that affect their relevance to human and ecological concerns? In Bacon, and almost continuously since, there has been talk of human dominion or control of the natural world. But in Bacon, there is also a love of nature and "pity for the sufferings of man,"[52] which has not always been so evident in the practice of technology.

Tension between these different paradigms extends right through the whole enterprise we call technology, from applied science and engineering to medicine. Often, it seems that the pyramid-building, control-asserting paradigm prevails, and "we remain in the dark ages, for power and the desire to control will not easily give themselves up to compassion and moral action."[53] But although political power (and economically oriented power relations) seem to determine this pessimistic conclusion, I hope to have shown that other, more personal aspects of life are significant, not least because of the considerable synergy that may develop between the personal and the political.

A third way of reflecting on the ethical meaning of technology, and the responsibilities that arise, might be to relate it to wider views (or intuitions) about the meaning of life. For some people, this statement may seem to imply the conventional assumption that ethics need to be grounded in religion, or else in some philosophical formulation that would be equally comprehensive. But I am making a different point, about the need to acknowledge human experience and understand the meanings arising from that. Of course, some relevant kinds of experience have traditionally been described in terms of religion. In reaction against that, many people dismiss the whole topic as too subjective to be worth discussion, and adopt positivist or reductionist philosophies. But such an attitude may too easily reinforce the trend toward object-centered habits of thought and dehumanizing applications of technology.

By contrast, when I encounter religious language, for example in discussion of an ethic of love (Chapter 7), or in talk about deep ecology (Chapter 6), my impulse is to dismiss any fundamentalist claims for its truth, but simultaneously, to seek to understand what aspects of personal experience and human sensitivity it reflects. Similarly, when I encounter mathematicians talking about eureka experiences (Chapter 2), or metal-workers of an earlier age using alchemical language, I want to understand what kinds of human feeling underlie their comments. Such questions affirm the significance of experience, and at the same time may enlarge one's ethical awareness.

Thus my way to understanding insists that the personal experience of individual people is real, important, and indeed, a source of ethical impulses (as when empathy is a check to violence). And I believe it important to go on insisting this in the face of reductionists, materialists, and others who think that all experience can be dealt with by psychological explanation, that is in turn reduced to a picture of biological mechanisms and genetic inheritance.

The biology and genetics involved may all be unassailably "true," but they do not represent the only point of view. We still need the perspective of the individual person with his or her inner life, and sense of meaning, purpose, and vitality in living. But at the same time, a proper commitment to the fullest possible use of rational thought ensures that philosophical accounts of experience, as well as psychology, biology, and genetics, all remain of central interest.

Poets and painters, musicians and writers of fiction often take a philosophy of this kind for granted to some extent. Among many such writers, Tolstoy stands out because he wrote with understanding of the agricultural technology of his day, even in works of fiction, and had some indirect influence on the later appropriate technology movement.[54] We have met one of Tolstoy's characters mowing grass in a hay meadow

(Chapter 1), testing a new threshing machine (Chapter 6), and discovering the meaning of parental love when his son was born (Chapter 7). All these experiences contributed to his sense of the meaning of life, but when he tried to define what it all added up to, he was baffled to find that it was "not to be put in words." Language tended to make the meaning of what he wished to say more elusive. Only when he stopped asking questions and got on with his work were the questions answered—"by life itself."[55]

According to playwright Christopher Fry, the bafflement encountered when one tries to explain any deep experience using words arises because ordinary language can express only what we have already mastered. If there are words to describe and express a thing, it is because people have already labelled and classified it. But music, he thinks, is more fundamental as a "universal language" through which we sense meaning in life and perceive unrealized possibility. For that reason, Fry wrote his plays in verse, using "sound and pattern related to music."[56] In technology, there is sound and pattern too, conveying a sense of purposiveness and potential. But we also have to define more specific purposes for practical work, and it is here that we are influenced by paradigms (which are often taken for granted), and by the more human impulses such as empathy or Baconian charity. Compartmentalized minds may sidetrack the latter. Yet to many individuals, sensibilities regarding people and nature seem central to what technology ought to be about.

Notes

Introduction

1. David E. Nye, *American Technological Sublime,* Cambridge: MIT Press, 1994, pp. 147–8, 199.

2. Ian Cox, *The South Bank Exhibition: A Guide,* London: HMSO, 1951, pp. 5, 62.

3. Langdon Winner, *The Whale and the Reactor,* Chicago: University of Chicago Press, 1986, pp. ix, 166–7.

4. Arnold Pacey, *The Maze of Ingenuity,* London: Allen Lane, 1974; 2d ed., Cambridge: MIT Press, 1992.

5. Arnold Pacey, *The Culture of Technology,* Cambridge: MIT Press, 1983.

6. David Dickson, *Alternative Technology and the Politics of Technical Change,* London: Fontana/Collins, 1974.

7. Winner, *The Whale and the Reactor,* pp. 21–2, 40. See also Winner's *Autonomous Technology,* Cambridge: MIT Press, 1977.

8. W. B. Bijker, Thomas P. Hughes, and Trever J. Pinch (eds.), *The Social Construction of Technological Systems,* Cambridge: MIT Press, 1987.

9. Stewart Russell, "The social construction of artefacts," *Social Studies of Science,* 16, 1986, pp. 331–346.

10. John Head, *The Personal Response to Science,* Cambridge: Cambridge University Press, 1985, p. 10.

11. Penny Williams, Toronto, private communication, 29 September 1997.

12. Michael Polanyi, *Personal Knowledge,* London: Routledge, 1958, pp. 67, 98, 130; Polanyi also uses the term "implicit knowledge," p. 52.

13. Gerald Holton, *Thematic Origins of Scientific Thought,* rev. ed., Cambridge: Harvard University Press, 1988, pp. 5–7.

14. Samuel C. Florman, *The Existential Pleasures of Engineering,* New York: St. Martin's Press, 1976.

15. I. Mitroff, T. Jacob, and E. T. Moore, "On the shoulders of the spouses of scientists," *Social Studies of Science, 7,* 1977, pp. 303–327.

16. Mary Midgley, *The Ethical Primate,* London: Routledge, 1996, pp. 14, 160.

17. Pacey, *The Culture of Technology,* p. 6.

18. Edward O. Wilson, *On Human Nature,* Harmondsworth, England: Penguin, 1995, p. 167.

19. Robert D. Romanyshyn, *Technology as Symptom and Dream,* London: Routledge, 1989, pp. 10–11.

20. Head, *The Personal Response to Science,* p. 31.

21. Merritt Roe Smith and Leo Marx, *Does Technology Drive History?* Cambridge: MIT Press, 1994.

22. H. H. Rosenbrock (ed.), *Designing Human-Centred Technology: A Cross-Disciplinary Project in Computer-Aided Manufacturing,* London: Springer-Verlag, 1989.

Chapter 1

1. Bruce Chatwin, *The Songlines,* London: Cape, 1987, chap. 3 (pp. 13, 16).

2. Rebecca West, *The Fountain Overflows,* London: Macmillan, 1957, pp. 305, 390.

3. R. Murray Schafer, *The Tuning of the World,* Philadelphia: University of Pennsylvania Press, 1980, pp. 83, 112. I am also indebted to Schafer for the comments on Honegger and Prokofiev, and others have pointed out an extensive use of railroad sounds in some forms of jazz. Philip Pacey (private communication, 1997) mentions many additional composers responding to technological sounds, including Hans Christian Lunbye, "Copenhagen Steam Railway Gallop" (1848); Charles-Valentin Alkan, "Le chemin de fer" (1844); Carlos Chávez, "Suite de Caballos de Vapor" (1926–7); Villa-Lobos, "The Little Train of the Caipira" (c.1930); John Adams, "Short Ride in a Fast Machine" (1986); and Jan Sandström, "Motorbike Concerto" (1986–9).

4. Marion Milner, *Eternity's Sunrise,* London: Virago, 1987, p. 83.

5. Paul Richards, "Agriculture as a performance," in Robert Chambers, Arnold Pacey, and Lori Ann Thrupp (eds.), *Farmer First,* London: Intermediate Technology Publications, 1989, p. 39.

6. Peter Pitseolak, quoted by Rudy Wiebe, *Playing Dead,* Edmonton (Alberta): NeWest, 1989, p. 47.

7. W. H. McNeill, *The Pursuit of Power: Technology, Armed Force, and Society,* Chicago: University of Chicago Press, 1982, pp. 126, 131.

8. Leo Tolstoy, *Anna Karenina,* trans. Constance Garnett, London: Heinemann, 1901, part 3, chap. 4, p. 248.

9. David Thomson, *Woodbrook*, Harmondsworth, England: Penguin, 1976, p. 191.

10. Anthony Storr, *Music and the Mind*, London: Harper-Collins, 1992, pp. 25–9.

11. John Bowlby, *Attachment and Loss*, vol. 1, *Attachment*, 1969; Harmondsworth, England: Penguin, 1984, p. 294.

12. Steve Backley, member of the British Olympic team, Barcelona, 1992, quoted by Frank Keating, "Backley first by a golden vision," *The Guardian*, 7 August 1992.

13. L. T. C. Rolt, *High Horse Riderless*, 1947, new ed., Bideford (England): Green Books, 1988, pp. 73–4.

14. David Joy and William T. Holt, "Improvements in hydraulic motive power engines," patent no. 2358, October 8, 1856. *Patent Abridgements, 32*, "Hydraulics," London: Eyre & Spottiswood, 1868, p. 706. On James Watt as an organ builder, see L. T. C. Rolt, *James Watt*, London: Batsford, 1962.

15. Percy A. Scholes, *The Oxford Companion to Music*, London: Oxford University Press, 9th ed., 1955, p. 276.

16. John Blacking, *How Musical is Man?* London: Faber, 1976, p. 116.

17. Thomas Hardy, *Tess of the D'Urbervilles* (1891), and D. H. Lawrence, *The Rainbow* (1915), quoted by Schafer, *The Tuning of the World*, pp. 73–6, 179.

18. I am indebted to Michael Bartholomew for information on motorcycles and on the history of the bicycle.

19. Schafer, *The Tuning of the World*, pp. 28, 83; also Carlo M. Cipolla, *European Culture and Overseas Expansion*, Harmondsworth, England: Penguin, 1970, p. 36.

20. Blacking, *How Musical is Man?* preface, pp. vi–vii.

21. Arthur Koestler, *The Sleepwalkers*, Harmondsworth, England: Penguin, 1964, pp. 398–9.

22. Peter Dronke, *Women Writers of the Middle Ages*, Cambridge: Cambridge University Press, 1984, pp. 197–8.

23. Galileo's work on vibrating strings and pendulums was in his book *Discorsi e Dimostrazioni Matematiche*, Leiden, the Netherlands, 1638; translated as *Two New Sciences*, New York: Macmillan, 1914; Dover ed., no date c.1955, pp. 95, 103.

24. Vincentio Galilei, *Dialogo di Vincentio Galilei . . . della musica antica et della moderna*, Florence, 1581, p. 3. I have also consulted two brief biographies of Vincentio's son, both entitled *Galileo Galilei*, by Ludovico Geymonat (New York: McGraw-Hill, 1965) and by Michael Sharratt (Oxford: Blackwell, 1994).

25. Leonard Shlain, *Art and Physics*, New York: Morrow, 1991, p. 277.

26. Douglas R. Hofstadter, *Gödel, Escher, Bach*, New York: Basic Books, 1979, pp. 552–5.

27. Storr, *Music and the Mind,* pp. 180–3. There are many other books on mathematics and music: See, for example, Edward Rothstein, *Emblems of Mind: The Inner Life of Music and Mathematics,* New York: Times Books, 1995.

28. Ian Stewart and Martin Golubitsky, *Fearful Symmetry,* Oxford: Blackwell, 1992, p. 259.

29. Gary Zukav, *The Dancing Wu Li Masters,* London: Rider/Hutchinson, 1979, p. 35.

30. Storr, *Music and the Mind,* pp. 146, 172–4; also Douglas Hofstadter, *Gödel, Escher, Bach,* pp. 626, 719.

31. W. H. Thorpe, *Bird-Song,* London: Cambridge University Press, 1961; also W. H. Thorpe, *Purpose in a World of Chance,* Oxford: Oxford University Press, 1978, pp. 47–51; Rosemary Jellis, *Bird Sounds and their Meaning,* London: BBC, 1977; and Laurie John, "The Music Lark: The work of David Hindley," *The Listener, 123,* 15 March 1990, pp. 8–9.

32. Joan Hall-Craggs, "The aesthetic content of bird songs," in R. A. Hinde (ed.), *Bird Vocalizations,* London: Cambridge University Press, 1969, p. 376.

33. For Hildegard, see Dronke, *Women Writers of the Middle Ages;* I am also indebted to BBC radio presentations. The other person quoted is Netsilik, an Inuit who features in Rudy Wiebe, *Playing Dead,* Edmonton Alberta: NeWest, 1989, p. 115.

34. Storr, *Music and the Mind,* p. 188.

35. Oliver Sacks, *A Leg to Stand On,* London: Duckworth, 1984; Picador ed., 1986, p. 87.

36. Sacks, *A Leg to Stand On,* pp. 108–9, 112.

37. Stewart and Golubitsky, *Fearful Symmetry,* pp. 72, 200.

38. Juliette Alvin, *Music Therapy,* new ed., London: Stainer and Bell, 1991, pp. 108, 112.

39. Samuel C. Florman, *The Existential Pleasures of Engineering,* New York: St Martin's Press, 1976.

40. Galilei, *The New Sciences,* Dover ed., p. 107.

41. These examples of "mechanical philosophy" come from the philosophers Descartes and Hobbes, and the physician Henry Power, and I have discussed them at greater length in Arnold Pacey, *The Maze of Ingenuity,* 2d ed., Cambridge: MIT Press, 1992, pp. 95–6.

42. Wilfrid Mellers, *Bach and the Dance of God,* London: Faber, 1980, pp. 8–9.

43. William Gilbert, *De Magnete,* London, 1600; for discussion of the context of this book and discussion of the beliefs of miners and alchemists, see David Goodman and Colin A. Russell (eds.), *The Rise of Scientific Europe, 1500–1800,* Sevenoaks (Kent), England: Hodder and Stoughton, 1991, pp. 145–156, 199–201.

44. Howard Rosenbrock, *Machines with a Purpose,* Oxford and New York: Oxford University Press, 1990, p. 155.

45. Rosenbrock, *Machines with a Purpose,* p. 155.

46. Rosenbrock, *Machines with a Purpose,* p. 176.

Chapter 2

1. See, for example, Margaret A. Boden (ed.), *Dimensions of Creativity,* Cambridge: MIT Press, 1996, referring to Anne Roe, pp. 135, 200, 212.

2. Anne Roe, "A study of imagery in research scientists," *Journal of Personality, 19,* 1950–51, pp. 459–470.

3. Roe, "A study of imagery." Also Anne Roe, *The Making of a Scientist,* New York: Dodd Mead, 1952; Anne Roe, *The Psychology of Occupations,* New York: John Wiley, 1956, pp. 76–7.

4. Quoted by Jacques Hadamard, *The Psychology of Invention in the Mathematical Field,* original ed., 1945; New York: Dover, 1954, pp. 142–3.

5. Hadamard, *Psychology of Invention,* pp. 83–5.

6. Eight engineers were initially included in one of Roe's studies. See Anne Roe, "A psychological study of physical scientists," *Genetic Psychology Monographs, 43,* 1951, pp. 121ff. For kinesthetic imagery, see Roe, "A study of imagery," p. 464.

7. Brooke Hindle, *Emulation and Invention,* New York: University of New York Press, 1981, pp. 49, 56, 83.

8. Eugene S. Ferguson, *Engineering and the Mind's Eye,* Cambridge: MIT Press, 1992, pp. 49–50.

9. W. Bernard Carlson, "Sketching as problem-solving," in Society for the History of Technology, *Abstract Document, Annual Meeting, 1–4 August 1996,* London: Science Museum, paper 9b, delivered 2 August 1996. Also W. B. Carlson and M. E. Gorman, "A cognitive framework to understand . . . Bell, Edison, and the telephone," in R. J. Weber and D. N. Perkins (eds), *Inventive Minds,* New York: Oxford University Press, 1992, pp. 48–79.

10. Stephen M. Kosslyn, *Image and Brain,* Cambridge: MIT Press, 1994, pp. 71–3; also Janice Glasgow et al. (eds.), *Diagrammatic Reasoning,* Cambridge: MIT Press, 1995.

11. Hindle, *Emulation and Invention,* pp. 93–5.

12. Gordon Fyfe and John Law (eds.), *Picturing Power: Visual Depiction and Social Relations,* London: Routledge, 1988, p. 3.

13. W. T. Stearn, quoted in Wilfrid Blunt, *The Compleat Naturalist: A Life of Linneaus,* London: Collins, 1971, p. 243.

14. Quoted by Arthur Koestler, *The Act of Creation,* London: Hutchinson, 1969, p. 170. See also articles by Nancy J. Nersessian and Gertrude M. Prescott,

in David Gooding and Frank James (eds.), *Faraday Rediscovered,* Basingstoke, England: Macmillan, 1985, pp. 15–19, 183–4.

15. Douglas Allchin, "The Q-cycle and . . . issues from the Ox-Phos controversy," unpublished paper presented at International Workshop on the Philosophy of Chemistry, Ilkley, England, July 1994. On some earlier models in chemistry, including graphic representation of chemical bonds, see C. A. Russell, *The History of Valency,* Leicester, England: Leicester University Press, 1971, pp. 90, 100–1.

16. Martin J. S. Rudwick, "The emergence of a visual language for geological science," *History of Science, 14,* 1976, pp. 149–195, especially 168–9.

17. Honor C. Farrell, *Gentlewomen of Science,* unpublished master's thesis, University of Leeds, 1994, p. 174; also Mott T. Green, *Geology in the Nineteenth Century,* Ithaca, NY: Cornell University Press, 1982, pp. 104, 123.

18. Ferguson, *Engineering and the Mind's Eye,* pp. xi, 137, 169–171.

19. Carl Mitcham, *Thinking Through Technology,* Chicago: University of Chicago Press, 1994, p. 224.

20. Thomas P. Hughes, *American Genesis,* Harmondsworth, England: Penguin, 1990, pp. 82–3.

21. Christopher Wood (ed.), *Dictionary of Victorian Painters,* Woodbridge (Suffolk), England: Antique Collectors' Club, 1971, 2nd ed., 1978, p. 338.

22. Farrell, *Gentlewomen of Science,* p. 126.

23. James Nasmyth, *An Autobiography,* ed. Samuel Smiles, London: John Murray, 1883, p. 125.

24. Eugene S. Ferguson, "The mind's eye: Nonverbal thought in technology," *Science, 197,* 1977 (26 August), pp. 827–836.

25. Arnold Pacey, *The Culture of Technology,* Cambridge: MIT Press, 1983, pp. 6, 49.

26. Leonard Shlain, *Art and Physics,* New York: William Morrow, 1991, pp. 103–118, 129; also Ian M. L. Hunter, *Memory,* Harmondsworth, England: Penguin, 1957, p. 143.

27. Gerald Holton, *Thematic Origins of Scientific Thought,* Cambridge: Harvard University Press, revised ed., 1988, pp. 385–7.

28. Nasmyth, *An Autobiography,* pp. 89–92.

29. Wolfgang Köhler, *The Mentality of Apes,* trans. Ella Winter, 1925; reprinted New York: Dover, 1973, pp. 32–3. For a more recent scientific study covering the wider issue, see Stephen Walker, *Animal Thought,* London: Routledge, 1983, pp. 382–8.

30. Roe, "A psychological study of physical scientists," pp. 182–7.

31. Anne Roe, "A psychological study of eminent psychologists and anthropologists, and a comparison with biological and physical scientists," *Psychological Monographs,* no. 352, Washington, DC: American Psychological Association, 1953, pp. 30–7, 50.

32. Roe, *Psychological Study of Eminent Psychologists*, pp. 50–1; also Liam Hudson and Bernadine Jacot, *The Way Men Think*, New Haven, CT: Yale University Press, 1991.

33. Margaret A. Shotton, *Computer Addiction? A Study of Computer Dependency*, London: Taylor and Francis, 1989, pp. 113–114.

34. Robert Chambers, *Whose Reality Counts?* London: Intermediate Technology Publications, 1997, pp. 36–8.

35. John Dunn, "Like father, like son. . .," *Professional Engineering*, 8 May 1996, pp. 14–15.

36. Simon Baron-Cohen, *Mind Blindness: An Essay on Autism and Theory of Mind*, Cambridge: MIT Press, 1995, pp. 32–48.

37. The agricultural researcher is discussed by Baron-Cohen, *Mind Blindness*, who quotes a previous discussion by Oliver Sacks. The comment about other scientists is my own observation, based partly on the discussion of Einstein's childhood by Holton, *Thematic Origins of Scientific Thought*.

38. Rudwick, "Emergence of a visual language." A series of articles that follows up Rudwick's pioneering study for biological sciences appeared in *Isis*, *84*, 1993, notably Alberto Cambrosio, Daniel Jacobi, and Peter Keating, "Ehrlich's beautiful pictures and . . . immunological imagery," pp. 662–699, and Jane R. Camerini, "Evolution, biogeography and maps," pp. 700–727.

39. The standard account of the invention of perspective is Samuel Y. Edgerton, Jr., *The Renaissance Rediscovery of Linear Perspective*, New York: Basic Books, 1975. However, my interpretation is based on an interest in Piero della Francesca. See, for example, Marilyn Aronberg Lavin, *Piero della Francesca: The Flagellation*, London: Allen Lane, 1972.

40. Arnold Pacey, *The Maze of Ingenuity*, 2d ed., Cambridge: MIT Press, 1992, pp. 50–2, 62–3.

41. Nicholas Copernicus, *De Revolutionibus Orbium Coelestium*, Nuremburg, 1543; trans. A. M. Duncan as *Copernicus on the Revolutions of the Heavenly Spheres*, New York: Barnes & Noble, 1976, p. 50; Copernicus also stressed symmetry in his preface, p. 25.

42. Brian Easlea, *Witch-Hunting, Magic and the New Philosophy of the Scientific Revolution*, London: Harvester, 1980, p. 59.

43. Alfred W. Crosby, *The Measure of Reality*, Cambridge: Cambridge University Press, 1997, p. 132.

44. The individuals referred to here are Francis Bacon, author of *The Advancement of Learning* (1605); Henry Power, author of *Experimental Philosophy* (1663); Henry Savile, author of an unpublished commentary on Ptolemy's astronomy and founder of Oxford University professorships of astronomy and geometry; Christopher Wren, one-time Savile Professor of Astronomy at Oxford; and Robert Hooke, author of *Micrographia* (1665).

45. Crosby, *The Measure of Reality*, pp. 163, 229.

46. Arthur Koestler, *The Sleepwalkers,* Harmondsworth, England: Penguin, 1964, p. 345.

47. Ferguson, *Engineering and the Mind's Eye,* pp. 169–177.

48. Shlain, *Art and Physics,* pp. 103–118, 129.

49. Sherry Turkle, *The Second Self: Computers and the Human Spirit,* London: Granada, 1984, p. 227.

50. H.-O. Peitgen and P. H. Richter, *The Beauty of Fractals,* Berlin and New York: Springer-Verlag, 1986, p. 3.

51. Roe, "A study of imagery," and Hadamard, *Psychology of Invention.*

52. Shotton (*Computer Addiction?*, pp. 44–5, 49) and Turkle (*The Second Self,* pp. 160, 206, 244) both comment on gendered behavior, mainly in the context of computing. See Roe, Psychological Study of Eminent Psychologists, on the "masculine image" of science.

53. Douglas Adams, *Dirk Gently's Holistic Detective Agency,* London: Heinemann, 1987, pp. 145–7.

Chapter 3

1. Gordon Glegg, *The Design of Design,* Cambridge: Cambridge University Press Engineering Series, 1969, pp. 18–19. On the eureka effect, see Arthur Koestler, *The Act of Creation,* London: Hutchinson, 1969, pp. 88, 106–7.

2. On Kekulé, see Koestler, *The Act of Creation,* p. 118; also Colin A. Russell, *The History of Valency,* Leicester, England: Leicester University Press, 1971, pp. 70, 242–3.

3. Henri Poincaré, *Science and Method,* trans. from French by Francis Maitland, London: Nelson, no date c.1914, pp. 53–4.

4. Eugene S. Ferguson, *Engineering and the Mind's Eye,* Cambridge: MIT Press, 1992, p. 169.

5. Charles Darwin, quoted by Mott T. Green, *Geology in the Nineteenth Century,* Ithaca, NY: Cornell University Press, 1982, p. 59.

6. Michael Polanyi, *Personal Knowledge,* London: Routledge, 1958, p. 101.

7. Polanyi, *Personal Knowledge,* pp. 52–3, 67, 88.

8. Douglas R. Hofstadter, *Gödel, Escher, Bach,* New York: Basic Books, 1979, pp. 674, 695.

9. Glegg, *The Design of Design,* pp. 18–20. See also Alliott Verdon-Roe, *The World of Wings,* London: Hurst and Blackett, 1954.

10. David P. Billington, *Robert Maillart's Bridges,* Princeton, NJ: Princeton University Press, 1979, pp. xiii, 98, 113–115.

11. David P. Billington, *The Tower and the Bridge,* New York, Basic Books, 1983, pp. 4, 6; also Glegg, *The Design of Design,* p. 1.

12. Evelyn Fox Keller, *A Feeling for the Organism: The Life and Work of Barbara McClintock,* New York: W. H. Freeman, 1983, pp. 69, 125, 149.

13. On responses to the new imaging technologies, see Michael Lynch and Samuel Y. Edgerton, "Aesthetics and digital image processing," in Gordon Fyfe and John Law (eds.), *Picturing Power: Visual Depiction and Social Relations,* London: Routledge, 1988, pp. 184–220.

14. James Hamilton-Paterson, *Seven Tenths: The Sea and its Thresholds,* London: Hutchinson, 1992, pp. 22–3, 29.

15. Tracy Kidder, *The Soul of a New Machine,* London: Allen Lane, 1982, pp. 34–5.

16. Cyril Stanley Smith, "Metallurgy as a human experience," *Metallurgical Transactions A,* 6, 1975, pp. 603–623; also Anders Lundgren, private communication, 2 August, 1996.

17. Boris Pasternak, *Doctor Zhivago,* trans. Max Hayward and Manya Harari, London: Collins/Fontana, 1961, chap. 14, p. 427.

18. David Thomson, *Woodbrook,* Harmondsworth, England: Penguin, 1976, pp. 174, 191.

19. George Sturt, *The Wheelwright's Shop,* Cambridge: Cambridge University Press, 1923; paperbook ed., 1963, pp. 54, 80.

20. Carl Mitcham, *Thinking Through Technology,* Chicago: University of Chicago Press, 1994, p. 220 (fig. 4).

21. Jade Snow Wong, *Fifth Chinese Daughter,* Harmondsworth, England: Penguin (Peacock ed.), 1965, p. 208.

22. Richard Gill, wheelwright of Beamsley, Yorkshire; private communication and lecture, 5 October 1995; also Sturt, *The Wheelwright's Shop,* p. 29.

23. Sturt, *The Wheelwright's Shop,* p. 45.

24. Smith, "Metallurgy as a human experience."

25. Robert M. Pirsig, *Zen and the Art of Motorcycle Maintenance,* London: Bodley Head, 1974, chap. 8.

26. Eric Lomax, *The Railway Man,* London: Cape, 1995, pp. 86–121.

27. Philippa Glanville, "Women silversmiths," *The Art Quarterly of the National Art Collections Fund,* 2, Summer 1990, pp. 24–29; also Philippa Glanville and J. Faulds Goldsborough, *Women Silversmiths,* London: Thames and Hudson, 1990 and Jane Lang, *Rebuilding St. Paul's,* London: Oxford University Press, 1956, p. 233.

28. Charles Webster, "Paracelsus confronts the saints," *Social History of Medicine,* 7, 1995, pp. 403–421.

29. Charles Webster, *From Paracelsus to Newton,* Cambridge: Cambridge University Press, 1982, pp. 1–12.

30. Webster, *From Paracelsus to Newton,* p. 4.

31. P. M. Rattansi, "Paracelsus and the Puritan Revolution," *Ambix, 11,* 1963, p. 24.

32. A. E. Waite (ed.), *The Hermetical and Alchemical Works of Paracelsus,* London, 1894, vol. 1, pp. 72–3.

33. Ahmad Y. al-Hassan and Donald R. Hill, *Islamic Technology,* Cambridge: Cambridge University Press and Paris: UNESCO, 1986, p. 133; the further quotation from Paracelsus in this paragraph can be found in A. E. Waite, *Hermetical and Alchemical Works of Paracelsus,* vol. 1, p. 167; vol. 2, p. 95.

34. John Webster, *Academiarum Examen, or the Examination of the Academies,* London, 1654, p. 106.

35. Webster, *Academiarum Examen,* p. 4.

36. Charles Webster, *From Paracelsus to Newton,* p. 71; also M. C. Gill, *The Yorkshire and Lancashire Lead Mines,* Sheffield, England: Northern Mines Research Society, 1986, pp. 6, 39.

37. Mark Slouka, *War of the Worlds,* New York: Basic Books, 1995, p. 11.

38. Jacques Ellul, *The Technological Society,* trans. John Wilkinson, London: Cape, 1965, p. 325.

39. Local information on the Sheffield steel industry, plus a comment by David Blunkett, quoted in *The Guardian,* 11 May 1989.

40. Shoshana Zuboff, "Informate the enterprise," in James I. Cash Jr., Robert G. Eccles, Nitin Nohra, and Richard L. Nolan (eds.), *Building the Information-Age Organization,* New York: R. D. Irwin, 1994, pp. 226–233.

41. Zuboff, "Informate the enterprise," p. 233; on "tacit knowledge," see Polanyi, *Personal Knowledge.*

42. Zuboff, "Informate the enterprise." Note that Zuboff's ideas about "informating" factories contrast sharply with older ideas about "automating" them, which would steadily reduce the amount of skill and knowledge required of operators. Compare Zuboff's views with the suggestions made by Rosenbrock, as quoted in the discussion of "skill-enhancing invention" in Chapter 9.

43. Morris Berman, *The Reenchantment of the World,* Ithaca, NY: Cornell University Press, 1981, p. 23.

Chapter 4

1. David E. Nye, *American Technological Sublime,* Cambridge: MIT Press, 1994, p. 335, note 15.

2. Raymond Williams, *Television: Technology and Cultural Form,* London: Fontana/Collins, 1974, pp. 10–19.

3. William J. Broad, *Star Warriors,* New York: Simon and Schuster, 1985.

4. George Monbiot, "Beware . . . mad scientist disease," *Guardian Weekly,* 10 December 1995, p. 12; for replies protesting the bias in Monbiot's col-

umn, see Neil Emans and Alexander Campbell on the letters page the following week.

5. Robert Chambers, *Whose Reality Counts?* London: Intermediate Technology Publications, 1997.

6. Eugene S. Ferguson, *Engineering and the Mind's Eye,* Cambridge: MIT Press, 1992, pp. 34–5.

7. This quality is evident in design drawings by locomotive builders, as illustrated by Ken Baynes and Francis Pugh, *The Art of the Engineer,* Guildford, England: Lutterworth Press, 1981.

8. Walt Whitman, "To a locomotive in winter," quoted in Kenneth Hopkins (ed.), *The Poetry of Railways,* London: Leslie Frewin, 1966, pp. 18–19. Nye, *American Technological Sublime,* p. 56, also quotes this poem.

9. Gwendolyn Wright, "Sweet and clean: The domestic landscape in the Progressive Era," *Landscape, 20,* 1976, part 1, pp. 38–43.

10. Ruth Schwartz Cowan, *More Work for Mother: The Ironies of Household Technology from the Open Hearth to the Microwave,* New York: Basic Books, 1983.

11. Sonia Livingstone, "The meaning of domestic technologies," in Roger Silverstone and Eric Hirsch (eds.), *Consuming Technologies,* London and New York: Routledge, 1992, pp. 117–123, 127.

12. Anne Karpf, "We have the technology . . .," *The Guardian,* 14 April 1992, reviewing Silverstone and Hirsch, *Consuming Technologies.*

13. Lewis Mumford, *The Myth of the Machine: Technics and Human Development,* London: Secker and Warburg, 1967, pp. 4–5, 140–7.

14. Fernande Saint-Martin, *The Semiotics of Visual Language,* Bloomington: Indiana University Press, 1990, p. xiv.

15. Susan Wittig, "The computer and the concept of text," *Computers and the Humanities, 11,* 1978, pp. 211–215.

16. Langdon Winner, "Artifact-ideas and political culture," *Whole Earth Review,* no. 73, Winter 1991, pp. 18–24.

17. Carroll Pursell, *White Heat: People and Technology,* London: BBC and Berkeley: University of California Press, 1994, p. 33.

18. W. B. Bijker, Thomas P. Hughes, and Trevor J. Pinch (eds.), *The Social Construction of Technological Systems,* Cambridge: MIT Press, 1987.

19. Bijker et al. (eds), *Social Construction of Technological Systems,* pp. 45–6.

20. Pierre Lemonnier, "On social representation of technology," in Ian Hodder (ed.), *The Meanings of Things,* London and New York: Harper & Row, 1989, pp. 161–8.

21. Michael Callon, "Society in the making," in Bijker *et al.* (eds.), *Social Construction of Technological Systems,* pp. 88–9.

22. Langdon Winner, *The Whale and the Reactor,* Chicago: University of Chicago Press, 1986, pp. 101, 169.

23. Pam Lin, "Microcomputers in education," in Tony Solomonides and Les Levidow (eds.), *Compulsive Technology,* London: Free Association Books, 1985, p. 94.

24. Nye, *American Technological Sublime,* p. xiii.

25. Nye, *American Technological Sublime,* pp. 278–9; see also pp. 38 and 59 on "manifest destiny."

26. Winner, *The Whale and the Reactor,* pp. ix-xi, 4, 14–15.

27. Winner, *"Artifact–ideas and political culture."*

28. Winner, *The Whale and the Reactor,* pp. 40, 54–8.

29. Winner, *"Artifact–ideas and political culture,"* especially p. 22.

30. Tom Athanasiou, "Artificial intelligence," in Tony Solomonides and Les Levidow (eds.), *Compulsive Technology,* London: Free Association Books, 1985, p. 26.

31. Thomas Crow, "Art and politics," *The Art Quarterly of the National Art Collections Fund,* no. 11, Autumn 1992, pp. 24–5.

32. Tony Solomonides and Les Levidow (eds.), *Compulsive Technology,* London: Free Association Books, 1985, pp. 5–6, 34, 94.

33. Carl Mitcham, *Thinking Through Technology,* Chicago: University of Chicago Press, 1994, p. 269.

34. Desmond Morris, *The Naked Ape,* London: Cape, 1967, p. 121. For a more critical discussion of the functions of play, see Caroline Loizos, "Play behaviour in higher primates," in Desmond Morris (ed.), *Primate Ethology,* London: Weidenfeld and Nicolson, 1967, pp. 176–218.

35. Jacob Bronowski, *The Ascent of Man,* London: BBC, 1980; also Konrad Lorenz, *Studies in Animal and Human Behaviour,* trans. Robert Martin, London: Methuen, 1970; Cambridge: Harvard University Press, 1970–71, vol. 2, pp. 234–5.

36. Paul Levinson, "Toy, mirror, and art," in Larry A. Hickman (ed.), *Technology as a Human Affair,* New York: McGraw Hill, 1990, pp. 294–304.

37. George Basalla, *The Evolution of Technology,* Cambridge: Cambridge University Press, 1988, pp. 7–10.

38. Basalla, *The Evolution of Technology,* pp. 66–70.

39. Peter Randall, *The Products of Binns Road,* London: New Cavendish Books, p. 5.

40. Iona and Peter Opie, *Children's Games in Street and Playground,* Oxford: Oxford University Press, 1969, p. vii; also Iona Opie, "Let's play skipping," *The Guardian,* 1, January 1991, p. 21.

41. John Nicholson, *Men and Women: How Different Are They?* Oxford: Oxford University Press, 1993, pp. 24–5, 108–9.

42. N. K. Humphrey, "The biological basis of collecting," *Human Nature,* February 1979, pp. 44–7.

43. David Noble, *America by Design: Science, Technology and the Rise of Corporate Capitalism,* New York: Oxford University Press/Galaxy Books, 1979, p. xxii.

44. Martin Pawley, "Johnson's journey into space," *The Guardian,* 1 December 1986, p. 11. For a comparable statement about critics of architecture, see Thomas A. Markus, *Buildings and Power,* London: Routledge, 1993, p. 26.

45. David Lodge, *Nice Work,* Harmondsworth, England: Penguin, 1989, p. 78.

46. William Golding, *The Spire,* London: Faber, 1965, chap. 8, p. 156; also compare chap. 6, p. 120.

47. Samuel C. Florman, *The Existential Pleasures of Engineering,* New York: St. Martin's Press, 1976, p. 125.

48. George Henderson, *Chartres,* Harmondsworth, England: Penguin, 1968, p. 14; also Arthur Koestler, *The Act of Creation,* 1964; Danube ed., London: Hutchinson, 1969, p. 296.

49. Mitcham, *Thinking Through Technology,* p. 247.

50. S. Körner, *Kant,* Harmondsworth, England: Penguin, 1955, pp. 181–2, discussing Immanuel Kant, *Critique of Judgement,* trans. J. C. Meredith, Oxford: Oxford University Press, 1911 and 1928.

Chapter 5

1. Wu Ch'eng-en, *Monkey,* trans. Arthur Waley, Harmondsworth, England: Penguin, 1961, pp. 84–5, referring in fact to a "finger of Buddha's hand."

2. Debra Rosenthal, *At the Heart of the Bomb,* Reading, MA: Addison-Wesley, 1990, p. 226.

3. Robert Chambers, *Whose Reality Counts?* London: Intermediate Technology Publications, 1997, pp. 22–7.

4. W. Stanley Jevons, *The Coal Question: An Inquiry Concerning the Progress of the Nation and the Probable Exhaustion of Our Coal Mines,* London and Cambridge, 1865; 3d ed., ed. A. W. Flux, London: Macmillan, 1906.

5. Donella H. Meadows, Dennis L. Meadows, Jørgen Randers, and William W. Behrens, *The Limits to Growth: A Report for the Club of Rome,* London: Earth Island, 1972.

6. Nicholas Georgescu-Roegen, *Energy and Economic Myths,* New York: Pergamon Press, 1976.

7. As this chapter was being completed, food disparagement laws in several U.S. states were in the news following a court case involving Oprah Winfrey, e.g., George Monbiot, "Give us this day," *Guardian Weekly,* 22 March 1998.

8. Charles Handy, *The Hungry Spirit,* London: Hutchinson, 1997, p. 28, quoting the "Index of Social Health" for the United States. For the worldwide "Human Development Index," see Mahbub ul Haq (ed.), *The Human Development Re-*

port 1990, Oxford and New York: University Press for the UNDP, 1990. New editions are published yearly.

9. Langdon Winner, *The Whale and the Reactor: A Search for Limits in an Age of High Technology,* Chicago: University of Chicago Press, 1986, pp. 169–174. Winner does not stress the term "quality of life," however.

10. Jane M. Howarth, "In praise of backyards: Towards a phenomenology of place," *Thingmount Working Papers on the Philosophy of Conservation, 96-06,* Lancaster, England: Department of Philosophy, University of Lancaster, 1996.

11. Howarth, "In praise of backyards," p. 16.

12. Morris Berman, *The Reenchantment of the World,* Ithaca, NY: Cornell University Press, 1981, pp. 15–17.

13. Edward O. Wilson, *The Diversity of Life,* Harmondsworth, England: Penguin, 1994, p. 5; also Evelyn Fox Keller, *A Feeling for the Organism: The Life and Work of Barbara McClintock,* New York: W. H. Freeman, 1983.

14. Geoffrey Cantor, *Michael Faraday: Sandemanian and Scientist,* Basingstoke, England: Macmillan, 1991, pp. 220–1.

15. Banks's Hunterian Oration, Royal College of Surgeons, quoted by Richard Hough, *Captain James Cook,* London: Hodder Stoughton, 1994, p. 48.

16. Berman, *The Reenchantment of the World,* pp. 124–131.

17. James Hamilton-Paterson, *Seven-Tenths: The Sea and Its Thresholds,* London: Hutchinson, 1992, p. 5.

18. Elizabeth Goudge, *The Joy of Snow: An Autobiography,* London: Hodder and Stoughton, 1974, pp. 111, 115.

19. John Clare, *The Shepherd's Calendar,* ed. Eric Robinson and Geoffrey Summerfield, London: Oxford University Press, 1964, p. viii and p. 3.

20. Flora Thompson, *Lark Rise to Candleford* (1945), Harmondsworth, England: Penguin, 1973; "Lark Rise," pp. 234–8.

21. Thomas Traherne, *Poems, Centuries and Three Thanksgivings,* ed. Anne Ridler, London: Oxford University Press, 1966, p. 177.

22. Debra Rosenthal identified some nuclear scientists who feel this way (though others clearly do not); see her *At the Heart of the Bomb,* pp. 87–8.

23. Gail Helgason, *The First Albertans: An Archaeological Search,* Edmonton, Alberta: Lone Pine, 1987, pp. 87–90.

24. Colin Tudge, *Last Animals at the Zoo,* Oxford: Oxford University Press, 1992, pp. 17–18.

25. Marcia Ascher, *Ethnomathematics,* Pacific Grove, CA: Brooks Cole, 1991, pp. 146–8; also, for more analysis of the visual thinking involved, see Edwin Hutchins, *Cognition in the Wild,* Cambridge: MIT Press, 1995, pp. 65–8; and for stick maps in the Marshall Islands, see Barbara E. Mundy, "Maps and mapmaking . . .," in Helaine Selin (ed.), *Encyclopaedia of the History of Science, Technology and Medicine in Non-Western Cultures,* Dordrecht, the Netherlands: Kluwer, 1997, pp. 587–591.

26. Ascher, *Ethnomathematics,* pp. 132–5.

27. Bruce Chatwin, *The Songlines,* London: Picador, 1988, chap. 2, pp. 119–20. There were other ways of mapping landscape also: See David Turnbull, "Maps and mapmaking of the Australian aboriginal peoples," in Helaine Selin (ed.), *Encyclopaedia,* pp. 560–2.

28. Keith H. Basso, "Stalking with stories," in Edward M. Bruner (ed.), *Text, Play and Story: 1983 Proceedings of the American Ethnological Society,* Prospect Heights, IL: Waveland Press, 1984, pp. 19–54.

29. Patricia Vinnicombe, "Rock art, territory and land rights," in Megan Biesele, Robert Gordon, and Richard Lee (eds.), *The Past and Future of !Kung Ethnography,* Hamburg, Germany: Helmut Buske Verlag, 1986, pp. 275–293, especially p. 290.

30. Ailton Krenak, "Hopes," in *Earth: A Supplement to The Guardian* (supplement marking the Earth Summit in Rio), London: Guardian Newspapers, 1992, p. 12.

31. Basso, "Stalking with stories," pp. 44–50.

32. Georges Duby, "Medieval agriculture," in Carlo M. Cipolla (ed.), *The Fontana Economic History of Europe, (1): The Middle Ages,* London: Collins/ Fontana, 1972, especially pp. 182, 195–200.

33. Hamilton-Paterson, *Seven-Tenths,* p. 221.

34. For a discussion of all the authors mentioned in this paragraph, see Wallace Stegner, *Where the Bluebird Sings,* New York: Random House, 1992, p. 205; for examples of Wendell Berry's essays, see his *Home Economics,* San Francisco: North Point Press, 1987.

35. Harold Horwood, *Dancing on the Shore: A Celebration of Life at Annapolis Basin,* Toronto: McClelland and Stewart, 1987, chap. 1.

36. Dennis Lee, "Cadence, country, silence: Writing in colonial space," *Boundary* (State University Press of New York), 2, 1974, pp. 151–168; also Margaret Atwood, *Survival: A Thematic Guide to Canadian Literature,* Toronto: Anansi Press, 1972, pp. 103–4, 124.

37. Robin Tanner, *Double Harness: An Autobiography,* London: Impact Books, 1990, p. 99; also John Constable, quoted by C. R. Leslie, *Memoirs of the Life of John Constable* (1843), London: John Lehmann, 1949, p. 104.

38. Hugh Brogan, *The Life of Arthur Ransome,* London: Hamish Hamilton, 1985, p. 3; also L. T. C. Rolt, *Landscape with Machines,* London: Longman, 1971.

39. Sonya Salamon, "Ethnic communities and the structure of agriculture," *Rural Sociology, 50,* 1985, pp. 323–340 (especially 338).

40. Joel A. Tarr, "From city to suburb," in Alexander B. Callow (ed.), *American Urban History: An Interpretive Reader,* New York: Oxford University Press, 1973, pp. 202–212; reprinted in Colin Chant (ed.), *Sources for the Study of Science, Technology and Everyday Life,* London: Hodder & Stoughton, 1988, vol. 2, pp. 28–37.

41. Richard Bradley, *Altering the Earth: The Origins of Monuments,* Edinburgh: Society of Antiquaries of Scotland, 1993, pp. 5–10, 17–20, 34–8. See also Vinnicombe, "Rock art, territory and land rights."

42. Lewis Mumford, *The Myth of the Machine: Technics and Human Development,* London: Secker & Warburg, 1967, p. 36.

43. Jane M. Howarth, "Neither use nor ornament: A conservationist's guide to care." *Thingmount Working Papers on the Philosophy of Conservation,* 96-05, Lancaster, England: Lancaster University Department of Philosophy, 1996.

44. Hamilton-Paterson, *Seven-Tenths;* Jane M. Howarth, "Nature's moods," *British Journal of Aesthetics,* 35, 1995, pp. 105–120.

45. Daniel R. Williams and Deborah S. Carr, "The sociocultural meanings of outdoor recreation places," in A. Ewert, W. Shavez, and A. Magill (eds.), *Culture, Conflict and Communication in the Wildland-Urban Interface,* Boulder, CO: Westview Press, 1993, pp. 209–219.

46. Basso, "Stalking with stories," p. 48.

47. Basso, "Stalking with stories," p. 48.

48. Freya Mathews, *The Ecological Self,* London: Routledge, 1991; compare Berman, *The Reenchantment of the World,* especially p. 237.

49. Mark Slouka, *War of the Worlds: Cyberspace and the High-Tech Assault on Reality,* New York: Basic Books, 1995, p. 69, also London: Abacus, 1996, p. 76.

50. Slouka, *War of the Worlds,* 1995 ed., pp. 71–2, 1996 ed., pp. 78–9.

51. Hamilton-Paterson, *Seven-Tenths,* p. 54.

Chapter 6

1. Thomas Jefferson, *Notes on the State of Virginia,* ed. William Peden, New York: W. W. Norton, 1972, pp. 164–5, 174.

2. Jane Howarth, "In praise of backyards," *Thingmount Working Papers on the Philosophy of Conservation,* 96-06, Lancaster, England: University of Lancaster Department of Philosophy, 1996, p. 12.

3. Mabel Shaw, *An Educational Venture in Northern Rhodesia,* London: Cargate Press,1948, pp. 188–9.

4. Denis Healey, *The Time of My Life,* Harmondsworth, England: Penguin, 1990, p. 12.

5. William Wordsworth, "The Prelude," Book I, lines 317–321, in *The Poetical Works of William Wordsworth,* ed. Thomas Hutchinson, London: Oxford University Press, 1923, pp. 636–7.

6. Philip Pacey, "Boys will be boys," unpublished essay discussing George Orwell's *Coming Up for Air* and William Wordsworth's "The Prelude," 1997.

7. Charles H. Gibbs-Smith, *Aviation: An Historical Survey,* London: HMSO, 2nd ed., 1985, pp. 1–9.

8. Alliott Verdon-Roe, *The World of Wings and Things,* London: Hurst and Blackett, 1954.

9. John Drinkwater, *The Outline of Literature,* London: Newnes, vol. 1, n.d., c.1925, p. 110.

10. Arthur Koestler, *The Act of Creation,* London: Hutchinson, 1969, pp. 353–8.

11. Aeschylus, *Prometheus Bound,* trans. James Sculley and C. J. Hetherington, London: Oxford University Press, 1975, p. 41, line 376. The best-known example of a historian of technology using the Prometheus myth is David S. Landes, *The Unbound Prometheus,* Cambridge: Cambridge University Press, 1969, see p. 24.

12. Koestler, *The Act of Creation,* pp. 353–8.

13. Daniel J. Boorstin, *The Discoverers,* London: Dent, 1984.

14. Richard Hough, *Captain James Cook,* London: Hodder & Stoughton, 1994, p. 250.

15. Boorstin, *The Discoveries,* p. 287.

16. Alan Moorhead, *The White Nile,* Harmondsworth, England: Penguin, 1973, pp. 28, 114, 135.

17. Francis Spufford, *I May Be Some Time: Ice and the English Imagination,* London: Faber, 1996, pp. 190–1.

18. George Malcolm Thomson, *The North-West Passage,* London: Secker & Warburg, 1975, p. 266.

19. Polly Toynbee, "Ran the world over," *The Guardian,* 15 October 1987, p. 13, reviewing Ranulph Fiennes, *Living Dangerously,* London: Macmillan 1987.

20. Ian Mackersey, *Jean Batten: The Garbo of the Skies,* London: Macdonald, 1990, pp. 279, 437.

21. James Hamilton-Paterson, *Seven-Tenths: The Sea and Its Thresholds,* London: Hutchinson, 1992, pp. 132–3.

22. Quoted by David Nye, *American Technological Sublime,* Cambridge: MIT Press, 1994, p. 192.

23. Jacques Ellul, *The Technological Society,* trans. John Wilkinson, London: Cape, 1965, p. 326.

24. Carl Mitcham, *Thinking Through Technology,* Chicago: University of Chicago Press, 1994, pp. 59–61.

25. Jeannette Page (ed.), *A Country Diary,* London: Fourth Estate, 1994, pp. 7–8; also Bill McKibben, *The End of Nature,* London: Viking, 1990, p. 54.

26. For an example of work by one of these artists, see *Andy Goldsworthy Sheepfolds,* London: Hue-Williams Fine Art, 1996.

27. Philip Pacey, *Charged Landscapes,* London: Enitharmon Press, 1978.

28. Peter Levi, *Collected Poems, 1955–1975,* London: Anvil Press, 1976, p. 66.

29. Hamilton-Paterson, *Seven-Tenths,* pp. 218–219; see also, on river valleys that invite roads or dams, Dixon Thompson, "Defining the frontier landscape," in Patrick A. Miller and Larry Diamond (eds.), *The Frontier Landscape: Selected Proceedings of the IFLA Congress 1981,* Vancouver: University of British Columbia, Landscape Architecture Program, 1982.

30. Nye, *American Technological Sublime,* pp. 124–5.

31. L. T. C. Rolt, *Landscape with Machines,* London: Longman, 1971, pp. 104–146.

32. James Nasmyth, *An Autobiography,* ed. Samuel Smiles, London: John Murray, 1883, pp. 163–5; discussed by L. T. C. Rolt, *Victorian Engineering,* London: Allen Lane, 1970, pp. 118–120.

33. Joseph Needham, with Wang Ling and Lu Gwei-Djen, *Science and Civilization in China,* vol. 4 (part 3), Cambridge: Cambridge University Press, 1971, pp. 229, 249, 295–6.

34. Francesca Bray, *Science and Civilization in China,* vol. 6 (part 2, *Agriculture*), ed. Joseph Needham, Cambridge: Cambridge University Press, 1984, p. 1.

35. Penny Jones, private communication, September 1994. Also, Lewis Mumford, *The Myth of the Machine: Technics and Human Development,* London: Secker & Warburg, 1967, pp. 143–4.

36. Andrew M. Watson, *Agricultural Innovation in the Early Islamic World,* Cambridge: Cambridge University Press, 1983, p. 117; also Donald Hill, *A History of Engineering,* Beckenham, England: Croom Helm, 1984, pp. 204–221.

37. Philip Pacey, "Landscape with trains," *16 mm Today* (railway modeling magazine), no. 63, February 1993, pp. 17–19; the schoolbook referred to is *Highroads of Geography,* London: Thomas Nelson, vol. 1, 1910 (no author acknowledged).

38. Quoted by Leo Marx, *The Machine in the Garden,* New York: Oxford University Press, 1964, pp. 102–5, 111–116.

39. Jefferson, *Notes on the State of Virginia,* pp. 164–5, 174.

40. Marx, *Machine in the Garden,* p. 127.

41. Frederick Turner, *Beyond Geography: The Western Spirit against the Wilderness,* New Brunswick, NJ: Rutgers University Press, 1992, p. 256.

42. Nye, *American Technological Sublime,* p. 133.

43. Leo Tolstoy, *Anna Karenina,* trans. Constance Garnett, new ed., London: Heinemann, 1972, pp. 761–3 (part 8, chap. 11).

44. David Thomson, *Woodbrook,* Harmondsworth, England: Penguin, 1976, p. 271. Compare Flora Thompson, *Lark Rise to Candleford,* Harmondsworth, England: Penguin, 1973, pp. 236–8 (quoted in chap. 5).

45. Gregory McIsaac and William R. Edwards (eds.), *Sustainable Agriculture in the American Midwest,* Urbana: University of Illinois Press, 1994, pp. 21–2.

46. McIsaac and Edwards, *Sustainable Agriculture,* pp. 45–6, 54–71; also Wendell Berry, *Home Economics,* San Francisco: North Point Press, 1987, pp. 162–178.

47. Henry David Thoreau, *The Maine Woods* (1864), ed. John J. Moldenshauer, Princeton, NJ: Princeton University Press, 1972, p. 155.

48. Michael Rothschild, *Bionomics,* New York: Henry Holt, 1990, pp. 273–282.

49. Michael E. Gorman, *Transforming Nature: Ethics, Invention and Discovery,* Boston: Kluwer Academic Publishers, 1998, pp. 240–5.

50. Jerry Mander, *In the Absence of the Sacred,* San Francisco: Sierra Club, 1991.

51. Colin Tudge, *Last Animals at the Zoo: How Mass Extinctions Can Be Stopped,* Oxford: Oxford University Press, 1992, pp. 14–28.

52. Freya Mathews, *The Ecological Self,* London: Routledge, 1994, pp. 140–1.

53. Alan Drengson, *The Practice of Technology,* Albany: State University of New York Press, 1995, p. 12–16.

54. Traditionally, the "affirmative way" pursued "perfection" through "delight in the created world." See Thomas Traherne, *Poems, Centuries, and Three Thanksgivings,* London: Oxford University Press, 1966, introduction by Anne Ridler, p. xvii.

55. Michael Pollan, *Second Nature: A Gardener's Education,* London: Bloomsbury, 1997, especially chap. 10, pp. 199 onward.

56. This information has been compiled from press reports, especially Paul Evans, "Pig haven in back gardens," *Guardian Weekly,* 1 February 1998.

57. Wilfrid Mellers, *Bach and the Dance of God,* London: Faber, 1980, pp. 298–302.

Chapter 7

1. Alan Paton, *Cry, the Beloved Country,* 1948, London: Cape, 1977, book 1, chap. 10, pp. 61–2.

2. Thomas Traherne, *Poems, Centuries and Three Thanksgivings,* ed. Anne Ridler, London: Oxford University Press, 1966, pp. 218, 243–4 (spelling modernized).

3. Elizabeth Goudge, *The Joy and the Snow: An Autobiography,* London: Hodder & Stoughton, 1974, p. 5; D. H. Lawrence, *Lady Chatterley's Lover,* 1928; Harmondsworth, England: Penguin, 1960, pp. 158–161.

4. Victor E. Frankl, *The Unheard Cry for Meaning,* New York: Simon and Schuster, 1978, pp. 24, 39.

5. H. H. Rosenbrock, "Engineers and the work that people do," *Control Systems Magazine (IEEE),* September 1981, pp. 4–8.

6. Eight engineers were initially invited to become part of the sample; Anne Roe, "A psychological study of physical scientists," *Genetic Psychology Monographs, 43,* 1951, pp. 121–239.

7. Liam Hudson, *Contrary Imaginations,* Harmondsworth, England: Penguin, 1967, pp. 50–2.

8. Sherry Turkle, *The Second Self: Computers and the Human Spirit,* New York: Simon and Schuster, 1984, p. 216; also, Margaret Shotton, *Computer Addiction?* London: Taylor & Francis, 1989, pp. 114, 166.

9. Turkle, *The Second Self,* pp. 201–2.

10. I am indebted to Dr. Carole Brooke (private communication) for the distinction between "doing things right" (efficiency) and "doing the right thing"; also Langdon Winner, *The Whale and the Reactor,* Chicago: University of Chicago Press, 1986, pp. x, 162–3.

11. Langdon Winner, *The Whale and the Reactor,* p. 51.

12. Robert M. Pirsig, *Zen and the Art of Motorcycle Maintenance* (1974), London: Vintage, 1989; Samuel C. Florman, *The Existential Pleasures of Engineering,* New York: St. Martin's Press, 1976.

13. Lewis Mumford, *The Myth of the Machine: Technics and Human Development,* New York: Harcourt, Brace, Jovanovich, 1967, p. 9; Martin Buber, *I and Thou,* trans. Ronald Gregor Smith, Edinburgh: T. & T. Clark, and New York: Scribner's Sons, 1958.

14. Herbert Read, *Art and Industry,* London: Faber, 5th ed., 1966, pp. 16, 187; Averil Stedeford, *Ellipse and Other Poems,* Oxford: Robin Waterfield, 1990, p. 11.

15. Francis Bacon, *The Advancement of Learning,* ed. W. A. Wright, Oxford: Clarendon Press, 1926, Book 1 (3), p. 7.

16. Judith Whyte, *Girls into Science and Technology,* London: Routledge, 1985, p. 91.

17. Francis Bacon, "The Masculine Birth of Time," in Benjamin Farrington, *The Philosophy of Francis Bacon,* Liverpool, England: Liverpool University Press, 1970, p. 62.

18. Carol Gilligan, *In a Different Voice,* Cambridge: Harvard University Press, 1982, pp. 64–7; also Mike W. Martin and Roland Schinzinger, *Ethics in Engineering,* New York: McGraw Hill, 2nd ed., 1989, pp. 18–20.

19. These statements may be queried since archaeological interpretations are colored by the gendered assumptions of archaeologists. However, see an attempt to use evidence from cave paintings to elucidate the gendered division of labor in prehistoric times by James C. Faris, "From form to content," in Dorothy K. Washburn (ed.), *Structure and Cognition in Art,* Cambridge: Cambridge University Press, 1983, pp. 90–111.

20. L. A. Moritz, *Grain Mills and Flour in Classical Antiquity,* Oxford: Clarendon Press, 1958, pp. 29–32, 134.

21. Mumford, *The Myth of the Machine: Technics and Human Development,* pp. 258–9.

22. Sayings of Jesus recorded in Oxyrhyncus Papyrus I, in M. R. James (ed.), *The Apocryphal New Testament,* London: Oxford University Press, 1924. This particular "saying" was also used in words of a hymn composed by Henry Van Dyke, one-time professor of English at Princeton. Also see Florman, *The Existential Pleasures of Engineering,* pp. 110–113.

23. Ingrid Palmer, "Seasonal dimensions of women's roles," in Robert Chambers, Richard Longhurst, and Arnold Pacey, *Seasonal Dimensions to Rural Poverty,* London: Francis Pinter, 1981, pp. 195–201.

24. Jane Brewer supplied metalwork for the west towers of St. Paul's Cathedral, London, in 1707, according to Jane Lang, *Rebuilding St. Paul's,* London: Oxford University Press, 1956, p. 233; Sarah Flower was a blacksmith recorded at Sonning, Berkshire, in the 1841 census (Earley District).

25. David B. Steinman and Sara Ruth Watson, *Bridges and Their Builders,* New York: Putnam, 1941, pp. 207, 246–7.

26. Alan Birch, *The Economic History of the British Iron and Steel Industry,* London: Frank Cass, 1967, pp. 11, 292–5.

27. Carroll Moore (producer) and Michael Billington (presenter), "Role Reversal," *Kaleidoscope,* BBC Radio 4, August 12, 1988. I am indebted to the BBC for a transcript of this program.

28. Liam Hudson, *Frames of Mind,* London: Methuen, 1968, p. 19; Brian Easlea, *Fathering the Unthinkable,* London: Pluto Press, 1983, pp. 65–8.

29. Easlea, *Fathering the Unthinkable,* pp. 45–8; for Mary Somerville, see Marina Benjamin (ed.), *Science and Sensibility: Gender and Scientific Enquiry, 1780–1945,* Oxford: Blackwell, 1991, pp. 49–52.

30. Philip Virgo, "Women into IT Campaign Report," February 1989, discussed by Peter Large and Neil Bradbury, "Skills Crisis in IT," *The Guardian,* 17 February 1989; also discussed by Carole Brooke, "Women and Information Technology," paper presented at European Women's Management Network Conference, Milan, February 1–2, 1991.

31. Cynthia Cockburn, "The circuit of technology," in Roger Silverstone and Eric Hirsch (eds.), *Consuming Technologies,* London: Routledge, 1992, pp. 32–44.

32. Farrington, *The Philosophy of Francis Bacon,* p. 59.

33. To try to answer this question, in 1986, I analyzed data from a sample of 68 women returning to jobs in computing, science, and engineering after having children. Many women's careers changed direction away from "hard" technologies, but there were often practical reasons for this rather than any marked change in attitude.

34. Ruut Veenhoven, "Is there an innate need for children?" *European Journal of Social Psychology,* 4, 1974, pp. 495–501.

35. Jane van Lawick-Goodall, "Mother-offspring relations," in Desmond Morris (ed.), *Primate Ethology,* London: Weidenfeld & Nicolson, 1967, pp. 287–346.

36. John Bowlby, *A Secure Base: Clinical Applications of Attachment Theory,* London: Routledge, 1988, pp. 85–8.

37. Ann M. Frodi and Michael E. Lamb, "Sex differences in responses to infants," *Child Development, 49,* 1978, pp. 1182–88.

38. Michael E. Lamb (ed.), *The Role of the Father in Child Development,* New York: John Wiley, 1976.

39. Many comments in this passage are derived directly or indirectly from the Women page of *The Guardian* in 1988–90; e.g., articles by Rosemary Burton (3 February 1988) and Wendy Webster (4 December 1990). On Kathleen Raine, see the piece by Naim Attallah on 23 March 1993.

40. Jane Dunn, *A Very Close Conspiracy: Vanessa Bell and Virginia Woolf,* London: Pimlico, 1991, pp. 163, 217, 223–227. Susan Hill is quoted from an informal comment, but see also her book, *Family,* London: Michael Joseph, 1989.

41. Margaret Drabble, *The Millstone,* Harmondsworth, England: Penguin, 1968, pp. 91, 102–3, 115.

42. Robert Bishop and Elizabeth Safanda, *Amish Quilts,* New York: Dutton, 1976; new ed., London: Laurence King, 1991, p. 14.

43. There are pictures like this with men dressed as harlequins by Picasso, but I am thinking of an interpretation including a mother and baby by John Nellist, to whom I am grateful for much discussion of the meanings of these pictures. See Jacques Lassigne, *Picasso,* London: Heinemann, 1955, plates 11, 13.

44. Easlea, *Fathering the Unthinkable,* pp. 10–15.

45. Judy C. Bryson, "Women and agriculture in sub-Sahara Africa," in Nici Nelson (ed.), *African Women in the Development Process,* London: Frank Cass, 1981, p. 37.

46. John Nicholson, *Men and Women: How Different Are They?* Oxford: Oxford University Press, new ed., 1993, p. 208.

47. Joan Smith, *Misogynies,* new ed., London: Vintage, 1996, pp. 153, 206.

48. Lawrence, *Lady Chatterley's Lover,* final chapter, p. 315.

49. Bowlby, *A Secure Base,* p. 6.

50. Leo Tolstoy, *Anna Karenina,* trans. Rochelle S. Townsend, London: Dent (Everyman ed.), 1928, part 7, chap. 16 (vol. 2, pp. 267–9).

51. Kate Figes, *Life After Birth,* London: Viking, 1998.

52. Dorothy Rowe, *The Courage to Live,* London: Harper-Collins, 1994, p. 285.

53. Francis Bacon, *The New Organon,* ed. Fulton H. Anderson, New York: Bobbs-Merrill, 1960, "The Great Instauration—Preface—Prayers," pp. 15–16.

54. Some ideas about God and the Big Bang are reviewed and dismissed by Paul Davies, *The Mind of God,* Harmondsworth, England: Penguin, 1992, pp. 58–61.

55. *The Bible,* King James (Authorized) Version, 1 John 4:7.

56. Goudge, *The Joy and the Snow,* pp. 17–18.

57. Esther Ehrman, *Mme. du Châtelet,* Leamington Spa, England: Berg, 1986, pp. 42–3.

58. Robert Reid, *Marie Curie,* London: Collins, 1974, pp. 92, 130.

59. Shelley Minden, "Patriarchal designs," in Patricia Spallone and Deborah Lynn Steinberg (eds.), *Made to Order: The Myth of Reproductive and Genetic Progress,* Oxford: Pergamon, 1987, pp. 102–6.

60. David Noble, "Automation madness," in Steven L. Goldman (ed.), *Science, Technology and Social Progress,* Bethlehem, PA: Lehigh University Press, 1989, pp. 75–6, quoting James Lighthill. See also Steven Levy, *Artificial Life: The Quest for a New Creation,* London: Cape, 1992.

61. David Lodge, *Nice Work,* Harmondsworth, England: Penguin, 1988, pp. 125–6.

62. Mary Shelley, *Frankenstein, or The Modern Prometheus,* ed. Marilyn Butler, London: William Pickering, 1993, p. 36.

63. Langdon Winner, *Autonomous Technology,* Cambridge: MIT Press, 1977, pp. 306–313.

64. David Knight, *Humphry Davy, Science and Power,* Oxford: Blackwell, 1992, pp. 121–2.

65. Benjamin, *Science and Sensibility,* pp. 45–7, 49.

66. Crosbie Smith, "Frankenstein and natural magic," in Steven Bann (ed.), *Frankenstein: Creation and Monstrosity,* London: Reaktion Books, 1994, pp. 39–59.

Chapter 8

1. Debra Rosenthal, *At the Heart of the Bomb,* Reading, MA: Addison-Wesley, 1990, pp. 33–37. Also quoted in this opening passage are informal comments by Jerry Ravetz, Maurice Barley, and Howard Rosenbrock (private communications, 1992, 1993, and 1997). See especially Jerry Ravetz, *The Merger of Knowledge with Power,* London and New York: Mansell, 1990.

2. David Elliott and Ruth Elliott, *The Control of Technology,* London and Winchester: Wykeham Science Series, 1976, pp. 92–8.

3. F. R. Jevons, *Science Observed,* London: Allen & Unwin, 1973, pp. 176–9. Scientists who pushed for ever more elaborate weapons were discussed by Solly Zuckerman under the title "Alchemists of the arms race," *New Scientist,* 21 January 1982, p. 188.

4. Cyril Stanley Smith, *A History of Metallography,* Chicago: University of Chicago Press, 1960, pp. 14–20, 41–9.

5. Rosenthal, *At the Heart of the Bomb,* pp. 56–7.

6. Rosenthal, *At the Heart of the Bomb*, p. 78.

7. Rosenthal, *At the Heart of the Bomb*, p. 34.

8. Noam Chomsky, *The Backroom Boys*, London: Fontana/Collins, 1973, p. 23.

9. Zygmunt Bauman, *Modernity and the Holocaust*, Cambridge: Polity Press, 1989, pp. 90–3.

10. Robert Jay Lifton, *The Nazi Doctors: Medical Killing and the Psychology of Genocide*, London: Macmillan, 1986, pp. 495–6.

11. Lifton, *The Nazi Doctors*, pp. 392–3, 446. The effect of routine was that "meaning came to lie in the performance of one's daily tasks rather than in the nature or impact of those tasks" (p. 459).

12. Albert Speer, *Inside the Third Reich*, trans R. and C. Winston, New York: Macmillan, 1970, pp. 212, 283, 524. Also quoted by Lifton, *The Nazi Doctors*, p. 494.

13. Bauman, *Modernity and the Holocaust*, pp. 195, 197.

14. Bauman, *Modernity and the Holocaust*, pp. 193–5.

15. Bauman, *Modernity and the Holocaust*, pp. 193–5, 198.

16. Robert Jay Lifton, *Home from the War: Vietnam Veterans neither Victims nor Executioners*, New York: Basic Books, 1984, p. 349; also Lifton, *The Nazi Doctors*, pp. 494–5.

17. Adrian Cullis and Arnold Pacey, *A Development Dialogue*, London: Intermediate Technology Publications, 1992, pp. 17–20, discussing John Lamphear, *The Traditional History of the Jie of Uganda*, Oxford: Clarendon Press, 1976.

18. Lifton, *The Nazi Doctors*, p. 159.

19. Lifton, *The Nazi Doctors*, pp. 159, 437n.

20. Lifton, *The Nazi Doctors*, p. 162.

21. Langdon Winner, "Technological frontiers and human integrity," in Steven L. Goldman (ed.), *Science, Technology and Social Progress*, Bethlehem, PA: Lehigh University Press, 1989, p. 50.

22. Thomas P. Hughes, *American Genesis*, New York: Viking Penguin, 1989, p. 3.

23. Vannevar Bush, *Science, the Endless Frontier*, Washington, DC: Government Printing Office, 1945; discussed by Don K. Price, *Government and Science*, New York: Oxford University Press, 1962, pp. 48–9.

24. Dennis Gabor, *Innovations: Scientific, Technological and Social*, New York: Oxford University Press, 1970, p. 9.

25. Robert Jungk, *Brighter than a Thousand Suns*, trans. James Cleugh, Harmondsworth, England: Penguin, 1960, p. 159 (chap. 11).

26. Thomas R. Berger, *Northern Frontier, Northern Homeland: Report of the Mackenzie Valley Pipeline Inquiry*, Ottawa, Ontario: Ministry of Supply and Services, 1977, vol. 1, p. 116.

27. David Noble, "Automation madness," in Steven L. Goldman (ed.), *Science, Technology and Social Progress*, Bethlehem, PA: Lehigh University Press, 1989, pp. 74–6.

28. Liam Hudson, "Research on intellectual types," *Advancement of Science* (British Association), 1966, pp. 612–616.

29. Tony Solomonides and Les Levidow (eds.), *Compulsive Technology: Computers as Culture*, London: Free Association Books, 1985.

30. Margaret A. Shotton, *Computer Addiction? A Study of Computer Dependency*, London: Taylor & Francis, 1989, pp. 5–6, 110–112, 166.

31. Sherry Turkle, *The Second Self: Computers and the Human Spirit*, London: Granada, 1984, pp. 212–215.

32. Tracy Kidder, *The Soul of a New Machine*, London: Allen Lane, 1982, pp. 50, 63.

33. Kidder, *Soul of a New Machine*, pp. 55–6, 90; Douglas Coupland, *Microserfs*, London: Flamingo, 1995. I am indebted to Penny Williams for a gloss on Coupland's novel and Kidder's journalism.

34. Turkle, *The Second Self*, p. 216.

35. On Nietzsche's "will-to-power" in the context of philosophy of technology, see Carl Mitcham, *Thinking Through Technology*, Chicago: University of Chicago Press, 1994, pp. 247–8.

36. E. M. Forster, *Howards End* (1910), Harmondsworth, England: Penguin, 1961, pp. 184, 198 (chaps. 23, 25); discussed by Earl G. Ingersoll, "*Howards End* and the engendering of the automobile," *Humanities and Technology Review*, 15, Fall 1996, pp. 11–24.

37. Anne Pivcevic (producer), "Car Crazy," BBC-2 Television, 19 March 1991.

38. Jungk, *Brighter than a Thousand Suns*, p. 183; Rosenthal, *At the Heart of the Bomb*, p. 62.

39. William J. Broad, *Star Warriors*, New York: Simon and Schuster, 1985, pp. 13, 57, 63.

40. David S. Landes, *The Unbound Prometheus: Technological Change and Industrial Development in Western Europe from 1750 to the Present*, Cambridge: Cambridge University Press, 1969, p. 24.

41. Christopher Marlowe, *Dr. Faustus*, lines 124, 126, in *The Works of Christopher Marlowe*, ed. C. F. Tucker Brooke, Oxford: Clarendon Press, 1969.

42. "The Hudson Report on Britain," Associated Business Programmes, 1974, quoted by Arnold Pacey, *The Culture of Technology*, Cambridge: MIT Press, 1983, p. 113.

43. Freeman J. Dyson, *Disturbing the Universe*, New York: Harper and Row, 1979, pp. 171–2.

44. Dyson, *Disturbing the Universe*, pp. 15–16; this aspect of Goethe is also quoted by Samuel C. Florman, *The Existential Pleasures of Engineering*, New York: St. Martin's Press, 1976, p. 145.

45. But see the discussion in J. R. Ravetz, *The Merger of Knowledge with Power*, London: Mansell, 1990, pp. 14, 69, 306.

46. Rosenthal, *At the Heart of the Bomb*, p. 34.

47. Noble, "Automation madness," pp. 74–6.

48. Shotton, *Computer Addiction?* p. 163.

49. Lifton, *The Nazi Doctors*, pp. 418–429.

50. Lifton, *The Nazi Doctors*, pp. 271, 276, 294 (on research equipment), and 453–5 (on long hours of medical work).

51. Lifton, *The Nazi Doctors*, pp. 401, 413–414.

52. *Goethe's Faust*, trans. Anna Swanwick, London: George Bell, 1905, scene 2, lines 764–5.

53. Oppenheimer's doubling is made clear both by Freeman Dyson, *Disturbing the Universe*, p. 87, and by Robert Jungk, *Brighter than a Thousand Suns*, pp. 129–144. Both authors also comment on Teller. Einstein was never part of the Manhattan Project, but his advice helped initiate it. Joseph Rotblat resigned when it became clear that Nazi Germany was not making an atomic bomb.

54. Thaddeus J. Trenn, "The central role of energy in Soddy's holistic and critical approach," *British Journal for the History of Science*, 12, 1979, pp. 261–276.

55. Freeman J. Dyson, "Bombs and poetry," in Sterling M. McMurrin, *The Tanner Lectures on Human Values*, Salt Lake City: University of Utah Press, vol. 4, 1983, pp. 81–146, especially 84.

56. Dyson, *Disturbing the Universe*, pp. 15–16, 171–2.

57. Alvin M. Weinberg, "Impact of large-scale science on the United States," *Science*, 134, 1961, p. 161.

58. Frank P. Davidson, *Macro: A Clear Vision of How Science and Technology Will Shape Our Future*, New York: William Morrow, 1983.

59. David E. Nye, *American Technological Sublime*, Cambridge: MIT Press, 1994.

60. Florman, *The Existential Pleasures of Engineering*; see also Florman's *Engineering and the Liberal Arts*, New York: McGraw Hill, 1980, p. 83.

61. Mike W. Martin and Roland Schinzinger, *Ethics in Engineering*, 2nd ed., New York: McGraw Hill, 1989, p. 304–5, quoting Florman, *The Existential Pleasures of Engineering*, pp. 121–2.

62. Joshua C. Taylor, *Futurism*, New York: Museum of Modern Art, 1961, pp. 124–6.

63. Lifton, *The Nazi Doctors*, pp. 129–30, 152.

64. Klaus Theweleit, *Male Fantasies*, trans. Stephen Conway, Chris Turner, and Erica Conway, St. Paul: University of Minnesota Press, 1987, 1989, vol. 2, pp. 197–9, quoting Ernst Jünger.

65. Theweleit, *Male Fantasies*, vol. 1, pp. 63, 79–81, 124–6, 215; vol. 2, pp. 179, 184.

66. Barbara Ehrenreich, in Theweleit, *Male Fantasies,* vol. 1, pp. ix–xvii.

67. Joan Smith, *Misogynies,* London: Vintage, 1996, pp. 36, 141–4, 175–7.

68. Broad, *Star Warriors,* pp. 24–5, 190.

69. Broad, *Star Warriors,* p. 51; also Peter Pringle and James Spigelman, *The Nuclear Barons,* London: Michael Joseph/Sphere (paperback), 1983, pp. 239–258.

70. Jane Goodall, *The Chimpanzees of Gombe: Patterns of Behavior,* Cambridge: Belknap Press of Harvard University Press, 1986, pp. 534, 593–4.

71. Frans de Waal, *Peacemaking among Primates,* Cambridge: Harvard University Press, 1989, pp. 171 (pointing out the closeness of humans to bonobos in terms of DNA), also 199–206. See also de Waal's later book, *Bonobo: The Forgotten Ape* (Berkeley and Los Angeles: University of California Press, 1997) and Richard Wrangham and Dale Peterson, *Demonic Males: Apes . . . and Human Values* (Houghton Miffin, 1996).

72. Barton C. Hacker and Sally L. Hacker, "Military institutions and the labor process," *Technology and Culture,* 28, 1987, pp. 743–775.

73. Robert D. Romanyshyn, *Technology as Symptom and Dream,* London and New York: Routledge, 1989.

74. Carol Bellamy, *The State of the World's Children 1996* (Annual Review of the United Nations Children's Fund), Oxford and New York: Oxford University Press, 1996, pp. 13–36.

75. Paul Richards, "Videos and violence on the periphery," *IDS Bulletin* (Institute of Development Studies, University of Sussex), 25, no. 2, April 1994, pp. 88–93.

76. Smith, *Misogynies,* p. 34.

77. Romanyshyn, *Technology as Symptom and Dream;* Carolyn Merchant, *The Death of Nature,* London: Wildwood House, 1980; Brian Easlea, *Fathering the Unthinkable,* London: Pluto Press, 1983.

78. Frederick Turner, *Beyond Geography: The Western Spirit against the Wilderness,* New Brunswick, NJ: Rutgers University Press, 1992, p. 24; also Romanyshyn, *Technology as Symptom and Dream,* p. 211; and Easlea, *Fathering the Unthinkable,* pp. 10–19.

79. The psychologists are Anne Roe, quoted throughout Chapter 2, and Margaret Shotton and Sherry Turkle, quoted toward the end of that chapter.

80. Romanyshyn, *Technology as Symptom and Dream,* p. 151.

81. Liam Hudson and Bernadine Jacot, *The Way Men Think,* New Haven, CT: Yale University Press, 1991.

82. George Steiner, *In Bluebeard's Castle,* London: Faber, 1971, p. 63.

83. Broad, *Star Warriors,* pp. 13, 104.

84. Patricia Werhane's concept, "moral imagination," is discussed by Michael E. Gorman, *Transforming Nature: Ethics, Invention and Discovery,* Boston, MA: Kluwer Academic Publishers, 1998, p. 191.

85. John le Carré, "New Horizons for Cold War Warrior," interview with David Streitfeld for *The Washington Post*, December 1996; reprinted, *Guardian Weekly,* 29 December 1996, p. 13.

Chapter 9

1. Langdon Winner, *The Whale and the Reactor,* Chicago: University of Chicago Press, 1986, pp. 61–83. Winner discusses AT in the United States during the 1970s, whereas most of my examples come from other parts of the world where AT continues to have influence.

2. E. F. Schumacher, *Small Is Beautiful: A Study of Economics as if People Mattered.* London: Blond and Briggs, 1973.

3. The agricultural engineer referred to was Adrian Cullis, and I edited his work. See Adrian Cullis and Arnold Pacey, *A Development Dialogue,* London: Intermediate Technology Publications, 1992.

4. Eugene S. Ferguson, *Engineering and the Mind's Eye.* Cambridge: MIT Press, 1992, p. 4.

5. Ivan Illich, *Tools for Conviviality,* London: Calder and Boyers, 1973.

6. J. Christopher Jones, *Design Methods: Seeds of Human Futures,* Chichester, England: John Wiley, 1970. (A new, rather different edition was published in 1997, but I am here drawing on work I did in the 1970s, using the first edition.)

7. Victor Papanek, *Design for the Real World,* London: Paladin, 1974.

8. E. F. Schumacher, *Good Work,* London: Abacus, 1980, pp. 3–4.

9. Mike W. Martin and Roland Schinzinger, *Ethics in Engineering,* New York: McGraw Hill, 2nd ed., 1989, pp. 335–6; also Papanek, *Design for the Real World,* p. 64.

10. Carole Brooke, *Journeys through the Quality Gap: Information Technology in Two Organizations,* Ph.D. diss. City University Business School, London, 1992.

11. Suzanne Gordon, "The importance of being nurses," *Technology Review,* October 1992, pp. 43–51.

12. Robert Jay Lifton, *The Nazi Doctors,* London: Macmillan, 1986.

13. Daniel A. Pollen, *Hannah's Heirs: The Quest for the Genetic Origins of Alzheimer's Disease,* New York: Oxford University Press, 1993, p. 238.

14. Chris W. Clegg, "Social systems that marginalize the psychological and organizational aspects of information technology," *Behaviour and Information Technology,* 12, 1993, pp. 261–6.

15. Carl Mitcham, *Thinking through Technology,* Chicago: University of Chicago Press, 1994, p. 134; quoting Thomas Kuhn, *The Structure of Scientific Revolutions,* Chicago: University of Chicago Press, 1970.

16. Robert Chambers, *Whose Reality Counts?* London: Intermediate Technology Publications, 1997, pp. 188–9, 193–4.

17. S. O. Funtowicz and J. R. Ravetz, *Global Environmental Issues and the Emergence of Second Order Science,* Luxembourg: European Commission, 1990; see also J. R. Ravetz, *The Merger of Knowledge with Power,* London and New York, Mansell, 1990, pp. 272–4.

18. H. H. Rosenbrock, "Interactive computing," Manchester, England: UMIST Control Systems Centre (report no. 338), 1977.

19. H. H. Rosenbrock, "Engineers and the work that people do," *IEEE Control Systems Magazine,* September 1981, pp. 4–8. This important paper has been reprinted several times, e.g., in Craig R. Littler (ed.), *The Experience of Work,* London: Heinemann, 1984, pp. 161–171.

20. Andrew Ure, *The Philosophy of Manufactures,* London: Charles Knight, 1835, pp. 1, 8, 23.

21. Howard Rosenbrock, *Machines with a Purpose,* Oxford: Oxford University Press, 1990, pp. 168–173.

22. Rosenbrock, "Engineers and the work that people do."

23. Charles Aspin and S. D. Chapman, *James Hargreaves and the Spinning Jenny,* Helmshore, England: Helmshore Historical Society, 1964; see also S. D. Chapman, *The Early Factory Masters,* Newton Abbott, England: David & Charles, 1967: Richard L. Hills, *Power in the Industrial Revolution,* Manchester, England: Manchester University Press, 1970.

24. Patricia J. Thompson, "A Hestian framework for science and technology," paper presented at the annual meeting of the Women's Studies Association, Minneapolis, Minnesota, 24 June 1988; see also Patricia J. Thompson, *Home Economics and Feminism,* Charlottetown: University of Prince Edward Island, 1988. (I am indebted to Ruth Carter for introducing me to Patricia Thompson's work.)

25. H. J. Mozans, *Woman in Science* (1913), new ed., Notre Dame, IN: University of Notre Dame Press, 1991, pp. 217–220. A memoir of Ellen Richards (1842–1911) was published in 1912 by Caroline Hunt.

26. Edward Atkinson, *The Science of Nutrition (including . . . The Aladdin Oven),* Springfield, MA: Clark W. Bryan & Co., 1892. (I am indebted to Sara E. Wermiel for a photocopy of this work.)

27. Thompson, "A Hestian framework."

28. Helen Appleton (ed.), *Do It Herself: Women and Technical Innovation,* London: Intermediate Technology Publications, 1995, pp. 6–7, 109–121.

29. Rachel Carson, *Silent Spring,* London: Hamish Hamilton, 1963 (British ed. published simultaneously with U.S. ed.).

30. Geoffrey Cantor, *Michael Faraday: Sandemanian and Scientist,* London: Macmillan, 1991, p. 220.

31. Max Jacobson, interview with Christopher Alexander, "Sector," *Architectural Design*, no. 12, 1971 pp. 768–770.

32. Christopher Alexander, Sara Ishikawa, and Murray Silverstein, *A Pattern of Language*, New York: Oxford University Press, 1977.

33. Freya Mathews, *The Ecological Self*, London: Routledge, 1994, pp. 126, 135–6, quoting also B. Devall and G. Sessions, *Deep Ecology: Living as if Nature Mattered*, Salt Lake City, UT: Peregrine Smith, 1985.

34. Cullis and Pacey, *A Development Dialogue*, pp. 93–5.

35. Freya Mathews, *The Ecological Self*, pp. 117–163.

36. Martha L. Crouch, "Debating the responsibilities of plant scientists in the decade of the environment," *The Plant Cell, 2*, 1990, pp. 275–7.

37. Arnold Pacey and Philip Payne (eds.), *Agricultural Development and Nutrition*, London: Hutchinson, 1985, pp. 150–9.

38. *The Plant Cell* is a journal of the American Society of Plant Physiologists. (For Crouch's article, see note 36.)

39. Phil Gates, "The environmental impact of genetically engineered crops," in *Biotechnology and Genetic Engineering Reviews*, vol. 13, ed. Michael P. Tombs, Andover, England: Intercept, 1996, pp. 181–195, especially 187, 189, 190.

40. Linda Lear, *Rachel Carson: Witness for Nature*, New York: Henry Holt, 1997.

41. I am thinking of some of my former students here, but Victor Papanek, *Design for the Real World*, is also a spokesman for this tendency.

42. Liam Hudson and Bernadine Jacot, *The Way Men Think*, New Haven, CT: Yale University Press, 1991, p. 99. (I tend to disagree with much that these authors say.)

43. Herbert F. York, *The Advisors: Oppenheimer, Teller and the Superbomb*, San Francisco: W. H. Freeman, 1976, pp. ix, 81. Other instances of men who were enthusiastic about physics or engineering when young, but who returned to more human concerns in later years, are discussed by Hudson and Jacot, *The Way Men Think*, pp. 99–105.

44. Edward O. Wilson, *On Human Nature*, Harmondsworth, England: Penguin, 1995, p. 106.

45. Charles Elliott, *Signs of the Times*, Basingstoke, England: Marshall Pickering, 1980, pp. 124–5.

46. Ray Wyre, "Pornography and sexual violence," in Catherine Itzin, *Pornography: Women, Violence and Civil Liberties*, Oxford: Oxford University Press, 1992, pp. 236–247.

47. John Nicholson, *Men and Women: How Different Are They?* Oxford: Oxford University Press, 1993, pp. 193–5.

48. Charles Moskos, "Battleground of confusion: The folly of comparing race and gender in the Army," *Guardian Weekly*, 18 January 1998, reprinted from the *Washington Post*.

49. Lewis Mumford, *The Myth of the Machine: Technics and Human Development*, London: Secker & Warburg, 1967, pp. 235–6.

50. Francis Bacon, *The New Organon*, ed. Fulton H. Anderson, New York: Bobbs-Merrill, 1960; see, first, the preface to "The Great Instauration," the prayers at the end, pp. 15–16; and second, Book I, Aphorism 129, on "dominion of the human race," alluding to Genesis 1:26–28.

51. Francis Bacon, *The Advancement of Learning,* ed. W. A. Wright, Oxford: Clarendon Press, 1926, Book I, 3 (p. 7).

52. J. R. Ravetz, *Scientific Knowledge and Its Social Problems*, Oxford: Oxford University Press, 1971, pp. 434–6.

53. I am indebted to Dr. Paul Gilbert of Derby for this phrase.

54. A Tolstoyan community in Leeds, later Pontefract, England, was founded soon after 1900 and manufactured bicycles in the 1920s and 1930s. Better documented, though, is the influence on AT via Gandhi, who corresponded with Tolstoy. Louis Fischer, *The Life of Mahatma Gandhi*, London: Granada, 1982, pp. 123–130.

55. Leo Tolstoy, *Anna Karenina,* trans. Constance Garnett, London: Heinemann, 1901, part 8, chap. 10 and 12, pp. 758, 764–5.

56. Christopher Fry, "Looking for a language," lecture delivered at the Ilkley Literature Festival, 19 September 1984; also Christopher Fry, *Plays: A Sleep of Prisoners . . . ,* London: Oxford University Press, 1971, p. 7.

Index